COSMOS AND CREATOR

By the same author

Les tendances nouvelles de l'ecclésiologie

The Relevance of Physics

Brain, Mind and Computers
(Lecomte du Noüy Prize, 1970)

The Paradox of Olbers' Paradox

The Milky Way: An Elusive Road for Science

Science and Creation: From Eternal Cycles to an Oscillating Universe

Planets and Planetarians: A History of Theories of the Origin of Planetary Systems

The Road of Science and the Ways to God
(Gifford Lectures, Edinburgh, 1975 and 1976)

The Origin of Science and the Science of its Origin
(Fremantle Lectures, Oxford, 1977)

* * * * *

The Ash Wednesday Supper (*Giordano Bruno*)
(translation with introduction and notes)

Cosmological Letters on the Arrangement of the World-Edifice (*J. H. Lambert*)
(translation with introduction and notes)

Universal Natural History and Theory of the Heavens (*I. Kant*)
(translation with introduction and notes)

Angels, Apes, and Men

COSMOS AND CREATOR

STANLEY L. JAKI

A Gateway Edition
Regnery Gateway

Printed in the United States of America

First published by Scottish Academic Press

This edition published by Regnery Gateway, Inc.
360 West Superior Street, Chicago, Illinois 60610

Library of Congress Catalog Card Number: 81-85511
International Standard Book Number: 0-89526-882-5

CONTENTS

PREFACE

This book would not have been written except for the urging of my esteemed friend, The Reverend Dr. Thomas F. Torrance, for many years Professor of Christian Dogmatics at the University of Edinburgh, and winner of the Templeton Prize for 1978. He felt that the major points made in my other, at times lengthy, books about creation, both as an article of Christian faith and as a foundation of natural science, should be made available in a concise form for the wider public. Those familiar with those books will not be exposed here to undue repetitions. A topic, so wide as to stretch across the cosmos and so deep as to cry out for the Creator, is ever fresh in perspectives and evidence. Whatever the prevailing norms of popularization, the issues dealt with here are too serious to allow dispensing with a modest measure of references and documentation. The same issues are also worth pondering with that seriousness which is not destined to be ever popular.

January 1980 S. L. J.

INTRODUCTION

A little over half a century ago Julian Huxley came forward with his *Religion without Revelation*, a book with a somewhat misleading title. What Huxley actually aimed at was to articulate a religion which not only was in no need of revelation but also excluded its very possibility by denying its rationality. The possibility and rationality of revelation rest on the existence of God, on his having made man to his own image – the image of his own rationality – and on his having placed him in a rationally coherent world which evidences its own and man's createdness. On the contrary, 'religion without revelation' called for a creed whose first tenet is a self-making man in a self-making universe. More radical challenge than this is hardly conceivable to the first article of the Creed: I believe in God, the Father Almighty, Maker of heaven and earth, of all things visible and invisible.

Such a challenge was not first posed by Huxley. He was merely an updated echo of a Renan, a Nietzsche, a Spencer, a Comte, and a Marx, relatively recent spokesmen of a tradition going much farther back in history. They all were at one in noting that nothing determines more decisively man's outlook on life than his facing up to the alternative: Existence, cosmic and human, is either dependent on a personal Creator or not. Needless to say, long after Huxley's writings had become dated, the tradition they represented kept producing further echoes. If an echo, usually a tease, comes with a fresh overtone, it can appear original enough to captivate the unwary. Jacques Monod's *Chance and Necessity* was certainly original as a title, but hardly as a message. All who before him did their best to exorcise the Creator from the scene, had no choice but to ascribe, in defiance of logic, everything to chance or/and necessity.

The most flagrant evidence of that defiance is the very activity which issues in a brilliant book or in a discovery worthy of a Nobel Prize. Neither books, nor scientific discoveries and awards can consistently be taken for a product of chance or of necessity, let alone of both. No wonder that in all such cases chance and necessity are furtively supplemented with that purpose and freedom which is certainly inherent in scientific research and philosophical reflection. The tactic is understandable, though wholly unjustified. It ushers in through the back door that common sense which gives meaning to any non-trivial, that is, well-reasoned use of the words chance and necessity. A religion which excludes the possibility of revelation cannot indeed be construed without falling back on such fairly transparent tactics. Their perennial popularity derives from man's longing for such a religion, hardly different from an aesthetic sentimentality, and from the sophistication of their construction and presentation. What makes some such tactics worth noting is the sense of urgency with which the self-anointed prophets of 'religion without revelation' reveal the vision of heaven on earth to each subsequent generation.

The crusading zeal of these prophets often mellows with age. When the new and revised edition of *Religion without Revelation* saw print in 1957, thirty years after the first, it did not contain the outspoken preface penned by Huxley in 1941, when his book was produced in an abridged inexpensive format* so that it might bring the good tidings to millions. As the Battle of Britain had by then been won, Huxley could set

* As No. 83 in *The Thinker's Library* (London: Watts & Co.). A second impression was issued in 1945. For the preface, see pp. v–viii.

himself to the task of planning for a new order of things. Its prospects were dependent, so Huxley declared, on meeting head-on and settling once and for all the alternative 'of God or not God, external Power or no external Power, non-human absolute values as against human evolving values'. Only after 'we had rid ourselves of Divine Power external to ourselves', Huxley argued, can we start thinking out 'the basis on which socially grounded humanist religion could develop healthily into something of value to evolving life, instead of as a menace to it'.

The most important part of that 'thinking out' was, according to Huxley, 'a retranslation of the realities at the base of orthodox Christianity into modern terms'. The most basic of such realities is, of course, the Creator of all. Therefore, in Huxley's words, the first definite statement of 'religion without revelation' had to be the claim that the Creator cannot be known, for man knows 'nothing beyond this world and natural experience'. The claim that 'man knows nothing beyond this world', is at least a tacit admission that man can know the world. If such is the case, 'religion without revelation' has the same basis as does religion with revelation or traditional Christian theism, to be specific. The latter, as was noted above, rests on knowing God, the invisible, from things visible, that is, the universe. This startling coincidence, in some respects at least, of the starting points of two diametrically opposite religions, with and without revelation, can help us to understand why so many nowadays speak of a creation of the universe and treat with a condescending smile the notion of Creator as personal God, the only source of Revelation.

As one may suspect, the inconsistency may only be apparent. The word creation can be used in such a diluted sense as to exculpate one of an inconsistency in not following it up with voicing the Creator. Huxley found that in his day 'many divines' earned applause by diluting the traditional idea of a personal God by pouring around it 'philosophical milk-and-water'. No different was the case with a 'number of scientists who have come to the conclusion that some sort of God exists somewhere in the background'. The sort of God thus made fashionable could, however, in Huxley's apt simile, produce no more than a 'faint cosmic smile' worthy of Alice's Cheshire Cat magnified to cosmic dimensions. Recently, some startling turns in scientific cosmology pushed to the foreground a misconceived notion of creation, and made the fashion of a spurious God or Creator stronger than ever. This fashion must first be seen for what it is, if a serious misunderstanding is to be avoided as to what is meant by the claim developed in the rest of this book that he who says cosmos must say Creator in the traditional sense if man's sense of reality, purpose, and consistency deserve more than lip-service.

CHAPTER ONE

AN UNEASY FASHION

Fashion new-fangled

In 1928, a year after *Religion without Revelation* first saw print, a powerful boost was given to a new-fangled, though also very uneasy, fashion of seeing in science a support of religion. The Gifford Lectures, which Eddington delivered between January and March 1927, came within the reach of the general public under the title, *The Nature of the Physical World.* It contained the memorable declaration: 'The idea of a universal Mind or Logos would be, I think, a fairly plausible inference from the present state of scientific theory; at least it is in harmony with it'. Much less remembered are the words that immediately followed: 'But if so, all our inquiry justifies us in asserting is a purely colourless pantheism. Science cannot tell whether the world-spirit is good or evil, and its halting argument for the existence of a God, might equally be turned into an argument for the existence of the Devil'.[1]

Julian Huxley must have been smiling, satisfied, and irritated. It must have been highly irritating to him that no less a scientist than Eddington had just joined the ranks of those who found science a pointer to the existence of some higher Power, perhaps a God. He must have felt satisfaction on finding his prognosis of the swelling rise of theological propensities among scientists verified by a fresh and outstanding case. Last, but not least, he could smile, and with a greater touch of reality than could Alice's Cheshire Cat, at the kind of God science was able to make popular and fashionable. Huxley must have known by then that the universal Logos, hardly a Creator, which Eddington's science allowed, was no more real than Eddington's cosmos. Already in 1920, in his *Space, Time and Gravitation*, a brilliant popularization of General Relativity, Eddington made no secret of the radically idealist philosopher he was as he declared in the way of a grand conclusion: 'We have found that where science has progressed the farthest, the mind has but regained from nature that which the mind has put into nature. We have found a strange foot-print on the shores of the unknown. We have devised profound theories, one after another, to account for its origin. At last, we have succeeded in reconstructing the creature that made the foot-print. And Lo! it is our own'.[2]

James Jeans, Eddington's great rival as first-rate science popularizer, was no less an idealist, a circumstance which hardly appeared significant to the millions of readers of his famed Rede Lectures, delivered in 1930 and published the same year under the title, *The Mysterious Universe.* It made a fad the notion that it was reasonable to think of a 'Great Architect of the Universe', provided one was thinking of him as a 'pure mathematician'.[3] Physical science had just turned so esoterically mathematical as to make respectable the permission given to God to appear on the scene, precisely because He could do so on the august pedestal of mathematics. Such a permission hardly meant a surrender to creation and Creator. Ironically, as will be discussed later, the death-knell was spelt on such a permission in the same year as it was given and published by Jeans. He died in 1946 without ever realizing that the Universe was mysterious, and in a far deeper sense than he suggested, because of mathematics.

Cosmos in embryo

No less revealing should it seem that in Eddington's and Jeans' success of making a fad (very superficially, to be sure) the inference to God from science, the most effective part was played by a factor, apparently not at all mathematical. The expansion of the universe, although with deep roots in the highly mathematical General Relativity, could be told with no mathematics and presented as an easily visualizable road to the very moment of creation. Experimental work which ultimately led to the recognition of the expansion of the universe, was started in 1913 by V. Slipher and greatly strengthened and systematized in the 1920s by Hubble and Humason. The theoretical work was started by de Sitter, who found Einstein's cosmological model to imply cosmic instability, that is, an expansion leading to infinite size and zero density. In 1922 Friedmann showed that under certain conditions the expansion can turn into contraction and that the two processes can alternate forever as if the universe were a spherical oscillator. Then in 1927 there came Lemaître's derivation from Einstein's cosmological model of the actually observed recessional velocity of galaxies, a derivation which for the first time tied theory and observation together.

Both observation and theory concerning the expansion of the universe were received by many with something akin to disbelief or at least with marked uneasiness. Einstein immediately perceived that the starting-point or moment of expansion as suggested by de Sitter's work represented something 'preferred', that is, a sharp departure from the equivalence of all positions and moments as demanded by relativity. In a letter of June 22, 1917, to de Sitter, Einstein vented his feelings with his customary candour: 'This point is thus *de facto* preferred . . . Naturally this does not constitute a disproof, but the circumstance irritates me'.[4] As to Friedmann's work, Einstein first slighted its merits, but by the early 1930s he warmed to it, possibly because it seemed to dissipate the vista of a 'preferred' initial moment in cosmic history. In the 'cosmic religion' professed by Einstein there was no room for creation or Creator.

Hubble, who as director of Mount Wilson Observatory controlled the observational work on the red shift of galaxies, could never bring himself to believing that the red shift in question meant recessional velocity. One could advance against this most obvious interpretation some reasonings which Jeans in his *Mysterious Universe* tried to put in their best light. He was clearly uneasy about the vision of a spectacular start some billion years ago of the cosmic process now under way. The vision it conjured up was best put by Lemaître: 'The evolution of the world may be compared to a display of fireworks that has just ended: a few red wisps, ashes and smoke. Standing on a cooled cinder, we see the slow fading of the suns, and we try to recall the vanished brilliance of the formation of worlds'.[5] Lemaître knew, of course, that the skies displayed enough brilliance even now. But compared with the celestial fireworks of a few billion years ago, the present activity of stars and galaxies could seem but a fading process. The universe was visibly ageing, a process which begins with a birth. In the case of an expanding universe, its ageing could not therefore fail to evoke the image of a compressed, embryonic cosmos, perhaps the moment of its very conception, or creation in time.

Neither in that article originally published in November 1931 under the title, 'L'expansion de l'espace', nor in some others on essentially the same theme, did Lemaître locate the creation of the universe in time at the beginning of its expansion. His reluctance to do so most probably had to do with the fact that as a well-trained priest he could hardly be unaware of Aquinas' view, which stood the test of time rather well; that neither the eternity, nor the temporality of the universe can be demonstrated on the basis of reason alone.[6]

One can only know from revelation, Aquinas argued, that the universe began in time. Of course, once the expansion of the universe was being traced backward, the impression of approaching an over-all starting point could seem overwhelming. In the Spring of 1934, in his Messenger Lectures given at Cornell University, Eddington described this impression in his habitually poignant diction: 'When some of us are so misguided as to try to get back milliards of years into the past, we find the sweepings piled up like a high wall, forming a boundary – a beginning in time – which we cannot climb over'[7] The boundary was the state in which all matter and energy constituting the universe were in their most organized and therefore most condensed state.

Eddington was not, however, to speak of creation let alone of a creation in time. As a scientist he was entirely within his right and responsibility to be far more concerned with the question 'whether the existing scheme of science is built on a foundation firm enough to stand the strain of extrapolation throughout all time and all space, than with prophecies of the ultimate destiny of material things or with arguments for admitting an act of Creation'. But could he justify his declaration that he had 'no philosophical axe to grind in this discussion'? He was certainly philosophical as he declared that he felt 'no instinctive shrinking from the conclusion' that the universe was heading into a final 'heat death'. He sounded even more philosophical as he qualified the idea of a cyclically self-rejuvenating universe to be 'wholly retrograde'. Most importantly, he explicitly rested his case on philosophy as he declared: 'Philosophically the notion of an abrupt beginning of the present order of Nature is repugnant to me, as I think it must be to most; and even those who would welcome a proof of the intervention of a Creator will probably consider that a single winding-up at some remote epoch is not really the kind of relation between God and his world that brings satisfaction to the mind'.[8]

Eddington felt the dilemma posed by the recession of galaxies, but was it an escape from that dilemma to imagine that the beginning was not the abruptness of a bang but a state in which 'there is no hurry for anything to begin to happen'?[9] Was this philosophical muddle, in which specious references to aesthetics smoothed over the radical abruptness of any real beginning, be it a bang or a whimper, not an inept cover-up of the very core of Eddington's philosophical if not theological convictions? Indeed it was, and the clue for this was given by Eddington himself. In speaking of that bang he declared already in 1928 in his famed Gifford Lectures that as a scientist he could not believe in it, and as a non-scientist 'he was equally unwilling to accept the implied discontinuity in the divine nature'.[10] Eddington was indeed a believer, but his God was that of pantheism, the usual landing-point of an idealist philosopher. On that landing he finds himself in the unexpected company of his materialist counterpart. Their difference is that of perspective not of essence, because both refuse to have recourse to the only factor, creation, which can make an essential difference. The difference between Nature seen by the pantheist as part of Divinity's nature and a Nature professed by the materialist to be the ultimate or divine, is a matter of perspective, not of essence, because both pantheists and materialists want to spare reality or nature from that discontinuity which is predicated by creation, but which alone, as will be seen, secures nature's continuity and consistency.

Uneasy panel

Around 1930 few equalled Eddington's stature as a witness of the fact that having to face up to the question of the creation of the universe under the impact of its expansion immediately became an uneasy fashion. The testimony of Eddington, who was most familiar with the scientific world of his day, is also priceless concerning the fact that many, perhaps most others, even some apparent theists, shared that uneasiness. In most cases it was visible only between the lines or at best indirectly alluded to. A classic illustration of this can be gathered from the brief lectures delivered by members of an illustrious panel organized for a discussion of the evolution of the universe as part of the centennial meeting of the British Association in London in late September 1931.[11]

The nine who spoke on the subject were Jeans, Millikan, Lemaître, Milne, de Sitter, Eddington, Lodge, Smuts, and Bishop Barnes. The latter, a skilled mathematician, was the *enfant terrible* of the Anglican Church of the day. Yet, if he now sounded terrible, it was not because he conjured up the vast vistas of cosmic and biological evolution. Nor was it in itself terrible that he expressed his firm belief in the existence of intelligent beings everywhere in the cosmos. But he claimed that radio communications were constantly reaching the earth from civilisations on planets around other stars, a terrible prospect because he foresaw mankind divided into two hostile camps over the agonizing problem whether those radio messages should be detected let alone be answered. Obviously, the Bishop, who had already heaped encomiums on Darwin, assumed that other intelligent species would be locked with us in a grim struggle for life with no holds barred.

Rather muted were the Bishop's encomiums for an entirely different outlook on the universe, an outlook embodied in Christian theism. True, the Bishop acknowledged that it had a strong and ennobling influence on European civilization through imparting the belief that the universe is due 'to creative thought and will, associated with purpose and plan'.[12] But about that plan he felt uncertain. What, he exclaimed, if science should prove that intelligence is merely a cosmic accident? Thus it was no accident that he admitted that he was 'by no means happy with regard to the expanding universe'. Without a firm belief in the once-and-for-all creation of all, a once-and-for-all expansion of all matter could easily appear, as it did to the Bishop, a mere cosmic accident, a rather dispiriting prospect. Was intelligence to be ultimately dissolved into the nothingness of an infinitely expanded matter? He found saving grace in the expansion insofar as it suggested a finite past for the universe. As a good mathematical physicist, he had well-justified reservations about infinitely large quantities, be they composed of bits of time or of matter. And, having a healthy zest for life, he was not in a mood to see through rose-coloured lenses the prospect of an endless cyclic sequence in which expansion alternated with contraction *ad infinitum* in the form of a cosmic treadmill.

The Bishop's theology was hardly a call for a philosophy ready to serve as a handmaid of verities transcending nature. As to Smuts, the spokesman of philosophy on the panel, he did not offer anything so clear as to be serviceable for a theology riveted on the createdness of nature. Holism and emergence, the two key words in Smuts' philosophy as well as in his address, proved themselves to be once more a smokescreen beyond which there was no advancing to a creation properly so called. The farthest Smuts was ready to go back in time and in reasoning was an 'undifferentiated primitive world-matrix which includes both the physical and the thought characters of the world'. It was in that matrix, and not beyond it, that Smuts located the source of Truth, Beauty, Goodness, and Love. They were for him as much 'structures of the evolutionary universe as the sun, the earth, and the moon'.[13]

Just as the uncertain theologian and the vague philosopher on that panel anticipated much of the fashion in which the question of origin was faced in the subsequent decades, so did the six scientists anticipate some apparently scientific slogans that have since become a fashion. Millikan chose to play the role of the hard-nosed experimentalist whose findings play ever new havoc with the best constructions of theorists prone to think that they have neatly fitted everything into their schemes. Thus Millikan could evade with apparent respectability the main topic and at the same time cast doubt on its reliability and significance. Instead of the origin of the universe, Millikan discussed the origin of cosmic rays. Were it to be proved, was the grand conclusion of his contribution, that cosmic rays can originate through annihilation (that is, complete conversion of matter into energy) not only in stars but in interstellar space as well, 'all future theories of the origin and destiny of the universe will be strongly influenced'.[14] Undoubtedly this was the case, especially if one was familiar with Millikan's ulterior motives to look for the origin of cosmic rays even in interstellar space as he was groping for what later became heralded as a steady-state universe. At any rate, in order to speak about the origin and destiny of the universe, one had to have a definite theory of the universe, a commodity hardly to be expected from that hard-nosed experimentalist which Millikan styled himself. He certainly could not take kindly to a theory of the universe which intimated an absolute beginning of it, as he had already made it clear that he was an eternalist.

Jeans seemed to be troubled by the conflict between the very long time demanded by stellar evolution, and the relatively short time offered by expansion. The latter, he claimed, 'brings almost complete chaos into the already chaotic problem of stellar evolution'. But all that Jeans' dicta illustrated was the chaos which a scientist can produce when he departs, however slightly, from his own speciality, in Jeans' case stellar evolution, gas theory, and planetary origins. He was not speaking with tongue-in-cheek as he attacked the idea of an expanding universe with phrases that were so many, almost incredible, blunders. He claimed that 'matter turns into energy and energy into mere bigness of space' as if space, a quarter of a century after Einstein's formulation of special relativity, had been a thing into which something else might conveniently turn. Worse, in blatant disregard of relativity, Jeans conjured up the future when 'galaxies will be running away from one another with speeds greater than that of light'. In that case no light, nor any other form of radiation could serve as a signal about the existence of galaxies. Only the mathematician, Jeans consoled himself and others, will be able to deduce their existence 'in recondite ways and probably no one will believe him'.[15] Whether anyone in the audience believed Jeans about space as something, or about speeds greater than light, is another matter. He certainly proved that anyone suspicious of an origin for the universe would end up in the not so original stance of parcelling the universe into wholly disconnected parts, island universes in the strict sense, as Jeans called those parts. He should have known that science always presupposed a coherent universe, the only kind of totality which provides meaning for its parts and for any particular question about them. Two such questions brought to a close Jeans' presentation and served as the starting point for Lemaître.

To Jeans' second question – whether the universe is expanding and at a rate indicated by the spectra of galaxies – Lemaître could point out that theory and observation called for an unequivocal yes. For Lemaître this meant that the universe formed originally a core which, as if it had been a primeval radioactive atom, broke into uncounted fragments, the elementary particles of the actual universe. Cosmic rays, and this was Lemaître's answer to Jeans' first question, were produced shortly afterwards when stars still had no atmosphere to block their escape. Lemaître's primeval atom was hardly an immediate success. He described cosmic evolution as starting with an original state in which the total mass of the universe 'would exist in the form of a unique atom; the radius of the universe, although not strictly zero, being relatively very small. The whole universe would be produced by the disintegration of this primeval atom'.[16] It still may turn out that the soup of a few kinds of elementary particles, which is taken today for the original form of the universe, is the product of the disintegration of Lemaître's primeval atom. However that may be, the present status of cosmology bears out the prophetic truth of Lemaître's remark: 'Cosmogony is atomic physics on a large scale'.[17]

A universe coming forth from a single gigantic atom could only appear a strikingly singular entity and prompt the question: Why is the universe what it is and not something else? The answer to such a question cannot be obtained from the universe itself, because the answer, as will be discussed later, would beg the question. Lemaître might have argued, and should have perhaps done so from that prominent forum, that modern cosmology or cosmogony not only unifies the largest realities, stars and galaxies, with the smallest, atoms and their parts, but also unifies science, an always restricted explanation, with metaphysics, the most unrestricted kind of explanation. Not, of course, with that pseudo-metaphysics which Smuts hailed to be embedded in quantum theory, but with a metaphysics which bears on physical reality and especially on its totality, the universe.

Perhaps Lemaître, a priest whose intense commitment to traditional Christian theism was an open secret, thought that the forum was not appropriate to serve as a sounding board for metaphysics. After all, he had just heard Eddington stop short of referring to creation and its goal-directedness. Were not both ideas in the wings as Eddington came forward with two startling items? One was his calculation of the value of the radius of the universe when its expansion started. The other was his remark that instead of billions (10^{12}) of years which, as he put it, 'have been fashionable of late' (Jeans having been the chief source of that fashion through his theory of stellar evolution), the cosmic past amounted to a mere ten thousand million (10^{10}) years. This relatively short cosmic past, Eddington warned, implied an important consequence 'in limiting the time available for evolution'.[18] Was a universe of so specific a size not custom-tailored by some external agent, a Creator, who also set its evolution on a very specific course? But let alone answering this question with a yes, Eddington did not even raise it.

Lemaître could not be unaware of the reason for Eddington's reticence and of the fashionable uneasiness behind it. But one can only guess his thoughts about Milne, a leading theoretician of cosmology and, unlike Eddington, a committed theist, who appeared to be intent on keeping the perspective of theism out of sight by referring at one point half-jokingly to the 'Celestial Plumber'.[19] Nor did some statements of de Sitter sound propitious for extending cosmology into metaphysics, a rather logical extension as will be seen later. 'The concept of the universe', declared de Sitter, 'as a closed entity is, so far as I can see a hypothesis, an arbitrary addition to and extension of the observed phenomena by our imagination'. With this declaration de Sitter came dangerously close to rejecting the idea of the universe as a limited totality of things, nay, the idea of totality itself. A curious risk for one of the most brilliant cosmologists of this century, but curious it certainly was that de Sitter, the renowned theorist, fell back into the role of the hard-nosed experimentalist. He claimed that any reference to whatever lies beyond our cosmic neighbourhood is a 'pure extrapolation'.[20]

To denounce the making of extrapolations beyond the 'cosmic neighbourhood' was to denounce the very soul of science and some of the most significant advances made throughout its history. Surprisingly, de Sitter did not feel guilty of making such 'pure extrapolation' as he, the only one on the panel to do so, pointedly recalled that the cosmic instability implied in Einstein's cosmological model permitted not only an expansion once and for all but also its alternation with contraction. In that case, the radius of the universe 'oscillates between zero and a finite maximum value in a rather short period'. He, however, admitted that the oscillating solutions 'require a density [of matter] exceeding that which is indicated by our knowledge of the distribution of matter in our neighborhood'.[21]

De Sitter must have perhaps stated that the actual density was almost a hundred times less than the one needed to turn the expansion into contraction. Perhaps he looked longingly, as did many after him, for that missing mass, a very fashionable longing for all those who want to exorcize the prospect of their createdness by trying to secure eternity to all matter through never ending cosmic oscillations. Revealingly, de Sitter was very explicit on one point, and here too he anticipated fashion: 'I do not propose', he stated, 'to discuss the question whether there ever was a beginning'. But this a-metaphysical and evasive statement was quickly covered up by the scientifically unimpeachable proviso: 'It suffices for my argument to define the "beginning" as that state of the universe and its constituent parts which we are with our present knowledge and theories content to use as a starting point, beyond which we do not wish, or are not able, to extend our investigations'.[22]

The difference between these two beginnings, one an

absolute origin meaning the creation of all, and another which refers to the condition and configurations of matter taken by science provisionally for the starting point of its investigations, is much too important to be dealt with in passing. The difference bears on the limitations of science which it has never been a fashion to keep in focus. Worse, when attention is focused on them, attendant circumstances can easily make the effort counter-productive. The role played by Lodge, then an octogenarian, at that panel is a case in point. He was well-known not only as a past President of the British Association, a recipient of coveted medals for scientific research, but also as a supporter for the previous fifty years of psychic research, even of the ways of communicating with the dead. Thus he could hardly sound convincing by pointing out a lack of basic clarification in the panel's programme. The lack in question related to the limitations of the scientific method, especially of the method of physics. While that method could cope brilliantly with the evolution and interaction of matter, it was woefully inadequate as soon as life and especially mind were also to be considered. Physics, Lodge stated, could in principle predict every ripple of the waves washing the seashore. But what if behind some ripples there was a fish, and behind some waves a boat with a rower in it? Could life and mind with a free-will be simply written-off because physics could not account for them? 'I venture to think', Lodge answered, 'that before we can philosophise upon such a theme as the ultimate fate of the universe, we must be able to take everything into account, and philosophise with a very wide and comprehensive knowledge of reality'. In that reality life and intelligence were for Lodge the far more important parts. To exclude them from the field of investigation was, as he put it, 'to philosophize from a restricted point of view [which] is interesting enough, but it is not conclusive'.[23]

Three attitudes

Participants of scientific panels, seminars, and congresses that were held in subsequent years in ever larger number on the origin, evolution, and destiny of the universe would have done well to recall and heed Lodge's advice. Instead, for the most part, as will be seen later, two attitudes dominated those gatherings. A third, too, was on occasion visible, but it never assumed a dominant role. According to the first, life and intelligence were inevitable results of the differentiation of physical reality. According to the second, whose most memorable spokesman was Teilhard de Chardin, all evolving physical reality was to be considered psychical. This was his remedy against what he called the antiscientific view which accorded the duality of mind and body. Rarely was a remedy more in need of cure. Indeed it was no less antiscientific than the physicalist reductionism of the first attitude.

In *The Phenomenon of Man,* where that remedy was proposed in a programmatic manner,[24] the most memorable detail is the Omega Point. Its theological content is clearly beyond the phenomenological framework to which Teilhard claimed to restrict himself. As is well known, that Omega was a point which Julian Huxley, who introduced the book to the English-speaking world, refused to accept as a rational and logical necessity. In this he could claim unwitting assistance from Teilhard's masterpiece. The cosmogony of the *Phenomenon of Man,* an enthralling poem in prose, was conspicuously void of any emphatic reference to creation and Creator. The end or Omega point clearly could not carry much conviction when Alpha the point of departure, or creation, had not been stated convincingly.

As to the third attitude, which is an unequivocal reassertion of the notion of creation in its traditional sense, it has made some momentous appearances, but none of them were free of some unnecessary burden. This too made uneasy the fashion of referring to the creation of the cosmos, though in a different sense than was the case with the first and second attitudes. In Milne's massive *Relativity, Gravitation and World Structure,* published in 1935, one could find, for instance, a special chapter entitled, 'Creation', which came to a close with truly eloquent phrases:

> 'The system to which we have likened the universe is an intelligible system. It contains no irrationalities save the one supreme irrationality of creation – an irrationality indeed to physics, but not necessarily to metaphysics . . . Those, who feel the question ["Why the universe?"] to be a permissible one, can legitimately answer the question "why?" by positing God. The physicist and cosmologist then need God only once, to ensure creation. For the biologist the world provides further opportunity for divine planning, if we admit the possibility of not entirely coincident evolutionary trends in similar circumstances. For man as more than cosmologist, as more than biologist, as possessing mind, possibly endowed with an immortal soul, God is perhaps needed always. Theoretical cosmology is but the starting-point for deeper philosophical inquiries'.[25]

Unfortunately, Milne's model of the universe embodied characteristics – extra-spatiality, extra-temporality, nay eternity, and perpetual youth (though without the cycles of rejuvenescence) – which, as Milne put it, are 'attributes we should like to associate with the nature of God'. Indeed, these are attributes, if not exclusively of God, of creatures known as pure spirits. To graft such attributes on the material creation, or cosmos, was to put an unnecessary burden on the notion and fashion of creation, a burden which Milne failed to alleviate.

Nothing directly metaphysical was contained in the burden which E. T. Whittaker produced by connecting scientific cosmology with creation, but burden it was and most debilitating in the long run. It consisted in his categorical declaration made at the very start of his Riddell Lectures given in 1942 that 'when by purely scientific methods we trace the development of the material universe backwards in time, we arrive ultimately at a critical state of affairs beyond which the laws of nature, as we know them, cannot have operated: a Creation in fact. Physics and astronomy can lead us through the past to the beginning of things, and show that there must have been a Creation'. The proviso, 'the laws of nature as we know them', obviously introduced into Whittaker's reasoning a *non sequitur* which was not offset by the truth of what he added in the same breath, namely, that 'of the Creation itself, science can give no account'.[26]

The *non sequitur* implied a crisis but in a sense not intended by Whittaker, as shown all too clearly by the very conclusion of his lectures. There he stated that by tracing backward the recession of nebulae the eyes of the scientist catch a glimpse of a 'crisis', that is, a moment sometime between one and ten billion years ago when the development of nebulae, stars, and planets began almost simultaneously. The true crisis of that statement was by no means caused by the uncertainty of the estimate of time to which Whittaker explicitly called attention. It rather lay in Whittaker's claim that physical science can lead to the beginning of all temporal processes, a claim which he reasserted by characterizing, without any further qualification, the beginning of the recession of nebulae as 'the ultimate point in physical science, the farthest glimpse that we can obtain of the material universe by our natural faculties'.[27] Brilliant historian of physics though he was, in addition to being a first-rate mathematical physicist (an almost unique combination), Whittaker was bafflingly oblivious to the fact that what produced a major crisis time and again in physics was to take a particular development of it for its ultimate stage. The logic of that crisis did not spare of its force Whittaker's reasoning. It prompted him to state that prior to that beginning or 'crisis' matter could only exist in a completely inert condition, a tempting inference but whose apparent reasonableness rested on a patently unreasonable premise, the unreasonableness of taking the physics of today for the physics of tomorrow. This is why Whittaker's reasoning provided no reasonable ground for the metaphysico-theological conclusion which postulated a creation and a Creator immediately antecedent to that allegedly ultimate state of affairs.

Four years later in 1946, Whittaker was still unaware of the inherent weakness of his reasoning. He flatly declared in his Donnellan Lectures given at Trinity College, Dublin, that 'different estimates converge to the conclusion that there was an epoch about 10^9 or 10^{10} years ago, on the further side of which the cosmos, if it existed at all, existed in some form totally different from anything known to us: so that it represents the ultimate limit of science. We may perhaps without impropriety refer to it as the Creation'.[28] The reference was most improper in the sense of not having that inner strength and consistency which alone assure that the point intended by the reference is indeed within its reach. Such an impropriety is a burden which plagued all subsequent endorsements of an apparently very scientific and therefore obviously most fashionable proof of the existence of God as construed by Whittaker. The most prominent of those endorsements came in an address of Pope Pius XII to the Pontifical Academy of Sciences on November 22, 1951. In justice to the Pope, his address, whose topic was 'The Proofs of the Existence of God in the Light of Modern Natural Science', was not a mere echo of what Whittaker wrote on the subject. But the Pope, who consulted with Whittaker in drafting his address, reproduced with full approval the passage from Whittaker's Donnellan Lectures quoted above and added: 'Thus with that concreteness which is characteristic of physical proofs, it [science] has confirmed the contingency of the universe and also the well-founded deduction as to the epoch [some five billion years ago] when the cosmos came forth from the hands of the Creator. Hence, creation took place in time. Therefore, there is a Creator. Therefore, God exists! Although it is neither explicit, nor complete, this is the reply we were awaiting from science, and which the present human generation is awaiting from it'.[29]

Needless to say, such a prominent endorsement of the view that physical science can reliably estimate the age of the universe (and not merely the past span of time of presently known physical processes) could only fuel the fashion of making a facile connection between cosmos and Creator. The Pope's address made headlines. Some saw in it a historical reconciliation between Science and Church as if there had ever been any fundamental conflict between the two, the case of Galileo notwithstanding. Others saw in it a sorely needed endorsement of science by ecclesiastical authority as if there was such a need. In many places it was eagerly studied as a document reliable and authoritative in every sense. The Pope's address, printed in many translations and study-editions, thus became that fashion which produces euphoria. A few were distinctly unhappy. One of them was none other than Lemaître, since 1952 the President of the Pontifical Academy of Sciences. As was already noted, he was not at all of the view that the existence of the physical universe prior to the beginning of its present expansion meant an absolutely inert and therefore uninvestigable condition.

Certainly unhappy with the Pope's address should have been the originators of the steady-state theory, although at least one of them, Hoyle, hardly cared less for what the Pope or any religious authority said on this or any other topic.[30] Whether the motivations of Bondi and Gold, the two other originators of the theory, were less tainted by the radically anti-religious sentiments to which Hoyle likes to give vent, is of no importance. The steady-state theory certainly aimed at eliminating the vista of an absolute beginning in cosmology. The feat, revealingly enough, was to be accomplished through the most glaring trick ever given scientific veneer, namely, the alleged coming into being out of nothing of one hydrogen atom per year in every cubic mile of space in order to keep the density of matter at a steady level while the galaxies receded from one another.[31]

The steady-state theory, which soon became for many a scientist a most respectable fashion, was a hapless philosophical and scientific somersault. Philosophically, as the Pope himself put it,[32] the theory was most gratuitous, a view also voiced from some rationalist quarters. What the proponents of the steady-state theory postulated most gratuitously (an altogether gentle characterization) was not the emergence of hydrogen atoms from radiation, a perfectly legitimate process, but their coming into being out of nothing and without a Creator![33] Apart from this, it is a purely gratuitous, or rather wholly fallacious, argument that if a specific excess is observed in the intensity of radiation produced by hydrogen atoms in cosmic spaces, it ought to be ascribed to hydrogen atoms coming into existence at a steady rate out of nothing, let alone independently of a Creator! Nevertheless, the steady-state theory became a respectable scientific fashion in evidence of the uneasiness of many with the notion of a genuine creation. In fact, in spite of all its intrinsic contradictions, the steady-state theory soon began to be spoken of as one of the three main rival theories of cosmology.[34]

The latest in fashion

Around the mid-1960s this rivalry ceased to be a triangular affair. Artificial satellites failed to detect the excess radiation postulated by the steady-state theory and the count of galaxies by radiotelescopes also dealt a major blow to it. While this was duly noted by the news media, the public remained for over a decade largely unaware of another development, although it was the most important cosmological feat since the finding, half a century ago, that the universe was expanding. That feat became a public sensation only when Arno Penzias and Robert Wilson received the Nobel Prize in 1978 for having detected around 1964 a cosmic radiation which on the one hand proved that the present expansion of the universe started at some ten billion years ago, and disproved, on the other hand, the claim of the steady-state theory that the expansion is a process under way since eternity at any point in an infinite universe. It was that radiation which, in proof of shallow journalism, was handed down to the public under such slogans as 'Echo of Creation'[35] and the like, slogans that certainly made a fashion.

As most fashions, the dubbing of the radiation discovered by Penzias and Wilson as the 'Echo of Creation', was a hollow proposition. That radiation, with its energy peak at 3°K, was, as will be shown later, no more an echo of Creation than any other specific phenomenon of cosmic range. It certainly strengthened the case for the so-called Big Bang theory, as the theory of the expanding universe has become known, which, rightly or wrongly, kept conjuring up the vista of an absolute beginning, a vista hardly to the liking of many a scientist. When these saw the sudden demise of the steady-state theory, they instinctively turned to the idea of an oscillating universe.

It began to serve as the only remaining rival to the Big Bang theory, though once again the rivalry was more a wishful thinking than a conflict of two theories with roughly equal merit. Actually, the allegation (scientific and journalistic) that such a rivalry existed only provided another proof of the strength of non-scientific (in this case clearly anti-metaphysical) motivations which can make even a shaky theory appear very reliable and popular.

What makes the theory of an oscillating universe so shaky relates, among other things, to the density of matter in the universe. Estimates of that density, largely based on the count of galaxies, keep yielding a value much too low in comparison with what is required if the gravitational attraction were ultimately to reduce to zero the recessional velocity of galaxies and pull them together again. When in early March, 1974, it was suggested that spiral galaxies contain far more than the missing matter needed to secure the oscillation of the universe, the jubilation of some scientists (and of almost the entire press and radio) was unmistakable.[36] It seemed as if the vista of creation had vanished once and for all and the spell of fashionable though uneasy references to it had been definitely broken. No wonder that when reports about that missing mass turned out to be wanting, the atmosphere in some circles became rather sullen. Walter Sullivan, chief science reporter of the *New York Times*, seemed to feel sympathy for E. R. Harrison, who unburdened himself in words that spoke for themselves. In likening the idea of a forever expanding universe to a march towards a 'graveyard of frozen darkness', Harrison noted: 'It would make the whole universe meaningless. If it were true, I would quit and spend my life raising roses'.[37]

There is a similar, clearly non-scientific, if not unscientific and anti-scientific, longing for an oscillating universe in S.

Weinberg's book, *The First Three Minutes*, through which the wider public began to be acquainted with the significance of the discovery of Penzias and Wilson, shortly before they were given the Nobel Prize. *The First Three Minutes*, an outstanding science popularization for the most part, conveyed covertly anti-religious and patently agnostic overtones. One wonders whether Genesis was not the real target as the Nordic saga of creation was held up by Weinberg to a gentle ridicule. Again, what was the cosmological or scientific justification of his conclusion that the study of scientific cosmology raises one's existence from the level of farce to that of tragedy?[38]

Just as hostility towards a theistic perspective of the cosmos can easily issue in hollow statements and transparent techniques, so can a covertly fundamentalist attachment to the perspective of creation. R. Jastrow, author of *God and the Astronomers*,[39] was clearly unaware of the wise reluctance of Maxwell, a most devout Christian, to go along with the suggestion that the existence of the ether, which no scientist doubted around 1870, evidenced the correctness of the biblical account which put the existence of light before that of the sun and the stars.[40] Jastrow and his publisher were possibly unaware of the fact that during the 1930s a memorable book with exactly the same title went through several reimpressions. It contained the text of the Warburton Lectures given in 1931–32 by William R. Inge, Dean of St. Paul's in London.[41] The brief text of Jastrow's book is, of course, no match to the informative richness of its much heavier namesake. Nor is it a match in this respect to the *First Three Minutes.* Indeed, it is no match at all to the critical eyes of a historian of science or of any reader ready to look up some quotations. Eddington is, for instance, patently misquoted by Jastrow.[42]

Thus it is with a certain misgiving that two other quotations given by Jastrow are reproduced here. Possibly, neither

Sandage nor Morrison spoke the words exactly as given by Jastrow and by the reviewers of his book who eagerly seized on them. Yet the substance of those words should seem credible. Sandage, who once described himself in my presence as a solipsist, and therefore hardly sympathetic to the idea of Creation, may have very well been taken aback by the growing evidence on behalf of the single expansion of the universe: 'It is such a strange conclusion . . . it cannot really be true'.[43] Philip Morrison, who in his book on Babbage, the father of modern computers, showed no sympathy for his hero's deep Christian faith,[44] reacted in much the same vein: 'I find it hard to accept the Big Bang theory; I would like to reject it'.[45] Sagan, the consummate artist of sentences resting on double, triple, and at times quadruple negatives, and a diligent detractor of historic Christian theism, must have felt uneasy in using the simple affirmative: 'The fires of creation are being observed today'.[46]

So much about an uneasy fashion. An unwittingly profound escape from its grip was sought by George Gamow in his *The Creation of the Universe*, a bestseller of the 1950s. Gamow, a strong advocate of the Big Bang, who even surmised something of the radiation discovered by Penzias and Wilson, hastened to warn the readers of the second printing of his book that he meant the word creation in the sense in which new fashions are being created in Paris.[47] Actually, a cosmologist can in a sense do far less with the cosmos than a fashion designer can do with a finely textured cloth. The latter can put his material in a wide variety of forms, though its texture always imposes some constraints on his freedom and fancy. The cosmologist is not even at liberty to order a material made for specification. He can only try to figure out the most convincing account of the cosmic texture. The cosmos and its structure are something radically given to the cosmologist and not produced or ordered by him in any sense. This consideration should seem to be far more meaningful to anyone looking for the Creator of the cosmos than the fragile though fashionable computations of its age. The much vaunted methods of science are not a whit more reliable for dating the moment of creation than the ones for which Bishop Ussher and others are still held up for ridicule and all too often by those who should rather reflect in some depth on what is really implied in their speciality, cosmology, or the study of that beauty which is the cosmos.

CHAPTER TWO

THE COSMOS OF SCIENCE

Cosmic Beauty

'The universe which science reveals to us is a dispiriting monotony. All the suns are drops of fire and all the planets drops of mud'.[1] So inveighed against science around the turn of the century Anatole France, the high priest of a literary apostolate the aim of which was, long before Julian Huxley, a religion without revelation. But why was France impelled to lay so heavy a charge at the door of science? As one who never did science, he felt that science could not supply man with the kind of inner fulfilment which only the contemplation of beauty could generate, and which alone seemed to him to be an effective substitute for the attractiveness of traditional Christianity.

Rarely did a charge go more to the heart of the matter, but was also more misplaced and more ill-timed. Without feeling, inspiration, and rapture life indeed is a dispiriting monotony. The most powerful factors which decide between hope and despair go far beyond the confines of cold reasoning and manipulation with numbers which can easily be taken for the genuine expressions of scientific activity. There was also some prima facie evidence behind France's characterization of science as an involvement which 'cares neither to please nor to displease'.[2]

If science was so foreign, as France thought, to that beauty which for us humans is the source of noblest pleasure, then one could readily take for a crude cosmetics the word, cosmology, or the study of the cosmos or universe. If the scientist must take for a drop of fire that sun which for a novelist is a flaming god, if the scientist must take for a drop of mud that earth which for the poet is a tender mother, is it still rightful to see in the scientific study of the physical world a cosmology, or a study of beauty in the most comprehensive sense? For such a beauty is meant by the word cosmos as coined by the Greeks of old. For them it stood for the universe as the most encompassing form of beauty. Of course, they had spoken of the world as cosmos long before they had begun its scientific study. But they did not feel that an injustice was done to the colourful beauty of heaven and earth as they discerned colourless, abstract, that is, scientific features in it.

Those features were geometrical and mathematical. Whatever of Anatole France's and other literati's insensitivity for such a beauty, it brought nothing less than rapture into the lives of not a few. Not all of them came close to the status of scientific genius. But first exposure to Euclid's *Elements* has repeatedly been an experience of rapture in the lives of those who, like Einstein, proved their genius through their scientific work.[3] To be sure, most of the many millions who had to go through the first book of the *Elements* in the upper grades of elementary school (without ever hearing the name of Euclid) felt at times the very opposite to rapture. At any rate, no feeling akin to rapture was felt by many of those millions who climbed the steps of the Acropolis to see the Parthenon. Much of its beauty resides in its being a subtle exercise in geometry. No different is the case with the beauty of pyramids, suspension bridges, and airplanes. All appear beautiful because they all embody in various ways proportion and symmetry which shine through almost all propositions of geometry.

Airplanes and bridges are particularly instructive in this respect. Pictorial representation of their development, which can be found in any good encyclopedia, shows that the more modern they are, that is, the more science they embody, the more beautiful they appear. Therefore the layman should beware of the urge to brush it off as sheer rhetoric when a scientist appraises the value of some proposition of theoretical physics on the basis of its 'beauty-content'. The value in question is the effectiveness of mathematical formulae to co-ordinate the known phenomena of nature and to provide increasingly wider control over them. A proposition of mathematical physics which has an added measure of beauty, say in the form of symmetry, can even disclose previously unsuspected phenomena of nature, a prerequisite for achieving control in greater depth. In looking back on his work, which first gave the startling glimpse of the existence of antimatter and perhaps of antiworlds, Dirac, one of the giants of modern physics, saw his achievement in a distinctly aesthetic perspective. According to Dirac, Schrödinger would have anticipated him had he had the courage to put more beauty into his famous equation by making it fully symmetrical.[4] Another work of Dirac, which made Maxwell's electromagnetic equations fully symmetrical, appears now to yield, in the discovery of magnetons, the counterparts of electrons, no less startling information about the universe.

The duality of matter and antimatter may conjure up some dark beauty, but so do all great dramas. If that beauty is dark, it is not only because of that annihilation, or rather instantaneous transformation, of matter and antimatter into radiation which takes place when they are not sufficiently removed from one another. The beauty in question is dark also because it carries man's eyes far beyond the visual range into the realms of the very small, particles and antiparticles, and of the very large, worlds and antiworlds. Much the same is the case with other major discoveries of modern physics. Today nothing is more trite than a reference to atoms and galaxies. A little over half a century ago it was still respectable to take atoms for mere hypotheses, and for most scientists there was only one galaxy. The thousands of nebulae visible through telescopes were taken to be relatively puny parts of our Milky Way.[5] No doubt, the universe as it appeared around 1900 had its own beauty which, however, pales into insignificance in comparison with what has since been unveiled by science about the universe.

Instead of one galaxy, science today counts billions of them, each with thousands of billions of stars. Instead of hypotheses about atoms, there is now a vast atomic science and technology. Photographs taken of distant galaxies turned into common knowledge a magnificent variety of patterns which are not a whit less beautiful than snowflakes and crystals under magnifying glasses. Atoms are still to be photographed, but models of atoms and molecules made of colourful balls connected with metal rods – staple features of introductory science courses – give a glimpse of the intricate beauty of the world of atoms. The beauty is not without its dark side, of which the mushroom clouds of atomic explosions are an awesome reminder. But its bright side has no less become an integral part of life. Countless jobs are involved in the production of an increasingly wider variety of synthetic materials, many of them with a texture and colour unattainable prior to the conquest by science of the realm of the atom.

The beauty of all this has a soaring quality in more than a poetical sense. Science reveals that matter, notoriously heavy, can literally take on wings. Thanks to solid state physics, components can be produced which make aviation history. On the one end of the gamut there is that transparent plastic, a mere five thousandth of an inch thick, which combines lightness and strength to such a startling extent as to make possible the flying over the Channel in a man-driven plane.[6] On the other end there is the enormous heat-resisting capacity of certain ceramics which assures the safe re-entry of space ships regardless of the metal-melting temperature developed by their searing re-entry through the atmosphere.

Unity and specificity

These two flights, one hugging close to the surface of earth and water, the other darting boldly toward the planets and stars, are symbolic of an attainment which puts modern science far above any previous stages of science and does so in a sense far deeper than what technological mastery over matter can convey. There was a time when it was generally believed that the earth's realm and the realm of the stars were ruled by essentially different laws. The realm of the earth, which was taken to reach almost as far as the moon's orbit, was that of ordinary matter in the form of four elements: earth, water, air, and fire. The other realm was that of the ether, a material which was believed to be the sole constituent of all celestial bodies and to be the reason for their circular motion. The ether was not only frictionless but also extremely rigid, a paradoxical combination which retained its curious credibility for a long time after Aristotle's cosmology and physics had been overthrown by Galileo and Newton.

Galileo, who held up Aristotle's ether for well-deserved ridicule, remained, however, a prisoner of the Aristotelian dichotomy between celestial and terrestrial motions by ascribing a necessarily circular motion to all celestial bodies.[7] Newton, who removed that dichotomy by showing that the motion of the moon was no different from that of a falling stone, failed in turn with respect to the Aristotelian dichotomy of matter. Newton retained the ether as a subtle medium, possibly the carrier of gravitational attraction and akin to space itself.[8] Thus the Newtonian universe was a universe saddled with a basic dichotomy, that between space (ether), which as

God's sensorium was infinite, and ordinary matter, which made up a finite universe of planets and stars. The infinity of ordinary matter became a widely entertained idea only during the 19th century, when the ether was still retained with no less contradictory qualities than was the case with Aristotle's ether. In order to be a carrier of light and other electromagnetic waves at their enormous speed, the ether had to be extremely rigid. It also had to be extremely tenuous so as not to present noticeable resistance to the enormous celestial bodies rushing through it with velocities which, though much less than the speed of light, far exceeded the velocity of the fastest bullets.

Not all late 19th-century physicists endorsed the contradictory qualities of the ether in the unhesitating manner of Lord Kelvin. For him it was enough to refer to the shoemaker's wax to resolve any and all doubts. Although fairly rigid, it presents no complete resistance to a piece of metal which, when placed on it, will sink through it at an almost imperceptible rate.[9] The ether, according to Lord Kelvin, combined those two qualities in a supreme degree. The real test for the ether did not lie in the teasing analogy of the shoemaker's wax. As a uniformly spread-out fluid, it had to modify the speed of light depending on its direction with respect to the observer. Extremely refined experiments were devised, such as the Michelson-Morley experiment, to detect such a modification, but the results were null. Still, not one of those who carried out those experiments saw in the results a proof of the non-existence of the ether.[10]

The demise of the ether came from a different direction which had to do with cosmic beauty in a far deeper sense than the experiments with the ether, beautifully conceived and designed as they could be. Although Einstein was aware of the repeated failures to detect the ether, his real concern was with Maxwell's electromagnetic equations.[11] They were possibly the most beautiful equations until then formulated in theoretical physics. As usual, beauty was not without problems of its own. Its most sensitive problem is to maintain itself whatever the surroundings and circumstances. But if light, a form of electromagnetic waves, was propagated in a medium at absolute rest, the type of rest synonymous with the ether, then those equations did not retain their form (beauty) if expressed in a frame of reference moving at steady velocity with respect to an observer assumed to be at rest. Unlike leading physicists of the time, young Einstein seemed to have been led by the conviction that by safeguarding the invariability of Maxwell's equations he saved their essential beauty and that this was of far greater importance than absolute rest and ether. Beneath his subconscious dissent from the prevailing preferences there lay a commitment to the beauty of the universe, a commitment with a distinctly metaphysical character of which Einstein became aware only years later.

In the eyes of many physicists Einstein's move emptied the universe not only of the ether but also of its intelligibility and unity which, in the form of a hallowed scientific dogma, had been equated for the previous 300 years with mechanical interaction.[12] Actually, Einstein's move made the genuine content of the term universe more unified than ever, and also far more intelligible and meaningful. Indeed, the first step in saving the beauty of Maxwell's equations was the unification of the speed of light by the postulate that its value remains in a vacuum always the same, regardless of the velocity of the emitting source, a postulate inconceivable within classical mechanics. An equally startling though logical consequence of Einstein's procedure was the unification of matter and energy. Their strict equivalence became a commonplace through the famous formula $E = mc^2$. Another, and no less important consequence was Einstein's almost immediate interest in extending his work from frames of reference that moved with respect to one another at any given velocity, to frames of reference accelerated with respect to one another. Such an acceleration was exemplified by the gravitational field produced by any mass. Unlike Special Relativity, General Relativity had therefore to have momentous cosmological consequences. They were spelled out by Einstein himself in 1917 in a paper in which for the first time cosmology came into its own, that is, achieved the status of a scientifically consistent discourse.[13]

Prior to that paper cosmology, which is the science of the totality of consistently interacting things (beauty, remember, always implies wholeness, a notion not far removed from wholesomeness), either set arbitrary limits to the universe, or when not setting such limits, the totality in question became burdened with self-defeating consequences. The closed Aristotelian universe, still a classic target of often uninformed ridicule, met its defeat in the dichotomy of its notion of matter as well as of motion. The originally finite Newtonian universe in infinite space was no less a precarious proposition.[14] Such a universe, unless given a specific rotation around its centre, must either collapse into a single mass or dissipate itself into infinity. A universe consisting of an infinite number of components (stars or galaxies) uniformly distributed in space is laden with a gravitational (and possibly also with an optical) paradox. Of such a universe there can be no scientifically consistent discourse, because in such a universe the gravitational potential is either zero or infinite at any point, a consequence which undermines the very existence of matter as we know it.[15]

The fact that the universe as pictured in Einstein's paper of 1917 was free of that paradox is not to suggest that the last word was spoken thereby in cosmology. Rather that paper was a first word to which an enormous wealth of scientific discourse has since been attached and in essentially the same sense, a fact suggestive of the ever stronger scientific conviction that there is a universe which is a consistent unity with very specific characteristics. This conviction was of the essence of Einstein's paper and of his subsequent cosmological work. The unity of the universe loomed large in Einstein's paper inasmuch as the total mass of the universe was taken by him to be finite. (He brushed off a possible solution of his cosmological equations with infinite mass as a mere foil that must not be taken seriously.[16]) The unity in question rests on the unity of the three dimensions of space and of the dimension of time. This four-dimensional manifold can only be treated within non-Euclidean geometries. Which of these would best serve scientific cosmology depends on the amount and distribution of matter constituting the universe. In Einstein's model the total amount was finite and produced a positive space-time curvature with the result that light or any material particle could travel in it only in closed orbits. More exactly, the total matter produced in that model a mesh of permissible paths of motion (such is the new definition of space) corresponding to a sphere. This is why there is no need to consider the question: what is beyond the confines of a universe with finite mass? Such a universe displays with overwhelming force the unity of all consistently interacting things and also a cosmic specificity or singularity best expressed in the space-time curvature produced by the totality of matter.

The correctness of this cosmological model is not related directly to the three most publicized experimental consequences of General Relativity: the advance of the perihelion of Mercury, the decrease of energy (frequency) of light (or the red-shift of its spectrum) when it escapes from a celestial body, and the gravitational bending of light. The model remained more of a mathematical construct than a reflection of cosmic reality until, as was noted in the previous chapter, it was tied through Lemaître's work to the recessional velocity of galaxies. It was through the further observation of that velocity that the idea of an expanding universe received its subsequent corroboration. Galaxies were found to obey with impressive regularity the law of expansion, that is, to have velocities proportional to their distances, as the farthest observed distance grew, partly through the advent of radio-astronomy, from a little over a hundred million light years to almost 200 times that distance, or twenty billion light years. This is the limit of the latest observations which show galaxies move with over half the speed of light.

Beauty unfolding

On the level of the very small, in the world of atom and nucleus, the triumph of the expanding universe was a far more complicated affair, both with respect to theory and observation, than the task of increasing the resolving power of telescopes, so that the recessional velocity of galaxies might be verified at ever greater distances. Yet progress in atomic physics was indispensable for a genuine grasp of the idea of the expanding universe, because galaxies in it are but a relatively late offspring of the evolution of atoms and molecules. Prior to the formation of galaxies, stars, and planets, the expanding universe had to be an expanding agglomerate of atoms and molecules which in turn was antedated by an expanding ball of even more elementary particles of matter. The problem of the expanding universe therefore soon appeared as the problem of the basic constitution and evolution of matter, an outcome which was a further proof of the impressive unity which modern science finds displayed throughout the universe. The pivotal aspect of the evolution of matter concerns the fact that in the actual universe almost three quarters of matter exist in the form of hydrogen and somewhat less than a quarter in the form of helium. All other elements constitute but a fragment of the total mass.

To account for that very specific distribution of matter among the hundred or so elements, it was necessary to construct a model of that primordial ball of elementary particles which would necessarily evolve into the present distribution of elements with a preponderance of hydrogen. Whatever the intentions of the American astronomer, H. Shapley, there was therefore a fortunate touch in his tasteless paraphrase of the opening words of St John's Gospel, 'In the beginning was the Word . . . and the Word was hydrogen gas',[17] because without hydrogen and its actual preponderance life, as we know it, would be impossible. About that time it began to be perceived that the condition of matter in that primordial ball had to be such as to produce a specific radiation. Theoretical work along these lines did not, however, become sufficiently accurate until the early 1960s. In 1964, the theorist P. J. E. Peebles concluded that there was possibly still present in cosmic spaces a radiation which marked the end of the first half a million years of the present expanding state of the universe. Without knowing of Peebles' work, and without even thinking of that primordial ball, Penzias and Wilson observed a radiation coming at the same intensity from every direction in space. It turned out to be the kind of radiation predicted by Peebles.[18] The radiation has the characteristics of the so-called black-body radiation which is the classic manifestation of an agglomerate being at least approximately in thermal equilibrium. The energy of such a radiation has its maximum at a specific frequency which in turn is an index of the temperature of the radiation. In the radiation as first observed by Penzias and Wilson that temperature is about 3°K, a temperature only 3 degrees above absolute zero, or 270°C below the temperature at which water freezes.

A radiation which marks the end of the first half a million years of the presently observed expansion hardly deserved the label 'an echo from the creation',[19] even if no time had preceded those half a million years. Indeed, as will be seen later, science is radically impotent to date the absolute start of physical processes, let alone the very first moment of the existence of matter. The real significance of the 3°K radiation, also called cosmic background radiation, lies in a far more meaningful aspect than in its apparent pointing to the moment of creation. First, the cosmic background radiation has discredited on strictly experimental grounds the steady-state theory in which such a radiation cannot occur. The cosmic background radiation is an irrefragable proof that the universe is not in a steady-state. Rather, the average temperature of the universe is steadily sinking, which is the very fact of cosmic expansion. Second, the cosmic background radiation assures enormous plausibility to the theoretical reconstruction of the cosmogonical process during the first half a million years of that expansion. Half a million years are but a fleeting moment when compared with the almost twenty thousand million, that is, twenty billion years, the presently observed time-span of expansion. But just as a plant is prefigured in its seed, the actual cosmos too is anticipated in that primordial gaseous ball of elementary particles. The analysis of that ball has the same advantage as the analysis of a seed. In the very small one can more easily grasp the striking specificity which may become almost elusive in the very large.

The analogy does not, however, hold in one very important respect. The growth of a plant from a seed is dependent on the constant accretion of material from the soil and air. The universe cannot contain more matter in its evolved state than it contained in its germ, that is, primordial gaseous ball form. Being the totality of consistently interacting things, the universe cannot add anything to itself nor can it lose anything. This is the cosmic variant of the principle of the conservation of matter, which is a principle of restriction, the most basic and encompassing quantum rule, so to speak. Within this principle of restriction there is an additional one. Although matter is equivalent to energy and although there is a transformation of matter into energy and vice-versa, the number of some heavy particles (baryons) remains essentially unchanged.[20] It is from those heavy particles (protons and neutrons) that the chemical elements are ultimately produced. The production, often called 'cosmic cooking',[21] has indeed a teasing analogy to ordinary cooking. As the latter, it too requires some 'minor' ingredients and a specific temperature. The desired cosmic outcome can be assured – and this shows something of the cosmic sweep of modern physics and of the utter coherence of the very large features of the universe with its smallest constituents – if one starts with a mixture which is at a thousand billion (10^{12}) degrees. (In the present state of fundamental particle physics it is rather problematic to go beyond that stage.) The mixture is composed of electrons, positrons, neutrinos, photons, or four kinds of light particles (leptons), and of one baryon (proton and neutron) for every billion (10^9) photons. Such is certainly a very specific 'cosmic soup' which hardly would encourage any sober mind into thinking that such is its only conceivable and necessary form. Such thinking is even more discouraged by every, more or less tentative, glimpse into states of matter, designated in the current literature as hadron, quark, and quantum phases of the cosmos, which may have preceded the cosmic soup described above. The original state of the universe, or the cosmic yolk, must therefore appear as a choice among an immense number of possibilities, a choice aimed at producing a most specific state of affairs: the cosmos in existence.

Once this specific cosmic soup is on hand, it begins to undergo, on account of its enormously high temperature and density, a rapid and effective cooking out of which there emerges our actual world of atoms, elements, molecules, galaxies and within these an enormous variety of stars, some of them perhaps with a planetary system. Within three minutes that cosmic soup cools to a still extremely hot one billion degrees, because it expands with a violent burst by millions of times its original volume. At this point its density is equal to the density of water, or more than a billion times less than its density at almost the very start of the expansion. About half an hour later the actual proportion of hydrogen and helium is on hand, that is, the very condition which permits the evolution of all matter into the actually observed universe in which the specific preponderance of hydrogen makes possible not only the burning of stars but also organic substances and life itself.

The next half a million years has its main interest in its being that phase of cosmic evolution in which all constituent particles are still in quasi-thermal equilibrium, including the primordial nebulae. In that quasi-equilibrium a black-body radiation is produced, but owing to the expansion, its energy and temperature are steadily decreasing. This also means that the frequency at which its energy is at its maximum shifts toward longer and longer wavelengths. After half a million or so years the radiation reaches its final stage which is still observable and was first detected by Penzias and Wilson. Clearly, on the scale of time, even if time had started only half a million years earlier, that radiation is not an echo of the *moment* of creation.

Still, the cosmic background radiation is a powerful pointer to the *act* or *fact* of creation, though for far more reliable reasons than the inept inference aimed at specifying its very moment. First, as was already noted, the cosmic background radiation discredits the steady-state theory, a construct forged with the implicit counter-metaphysical aim[22] to secure for the universe along the parameter of time that infinity which it has lost with respect to the quantity of matter constituting it. The eternity of the universe would then, according to a very old scenario, make plausible the uncreatedness of the universe. Of course, even the steady-state theorists could not argue that the rate of expansion of the universe (the recessional velocity of galaxies) had necessarily to be what it actually is, a precondition to the assertion that the emergence of new hydrogen atoms (whatever their being created out of nothing and without a Creator) had to be necessarily such as postulated by them. Nor could the steady-state theorists argue that matter had to exist necessarily in the form of hydrogen and its components, electrons, protons, and neutrons. But however specific the universe still was within the steady-state theory, it was always the same and eternal.[23] As such it could blind many minds (wizards though they may be in mathematical physics) to the fact that anything specific, that is, peculiarly limited to a particular form, is hardly the only and necessary form of existence.

Second, the cosmic background radiation forcefully evidences a specific cosmic evolution dependent on conditions restricted to within extremely narrow limits. One of them is the original ratio of photons to protons, neutrons, and electrons. If that ratio had been slightly less than the value given above, much perhaps all hydrogen would have turned into helium. In that case the universe would have been deprived of all organic life. Another of those conditions relates to the total number of those particles, that is, to the total mass of the universe. If that mass had been slightly more than indicated by the actual rate of expansion and other observations, the expansion would have, on account of the greater gravitational attraction, been too slow to permit the cosmic cooking of elements which must produce them also in their actual, most peculiar relative percentages. With markedly more matter originally present, the expansion (and the cosmic cooking) would not have taken place at all. Had that mass been slightly less, the expansion would have been much too fast to maintain temperatures and pressures necessary for that cosmic cooking. The universe can indeed be said to have had a very narrow escape in order to become what it actually is. Indeed it may be said that the universe weighs as much as it does, because we humans are here: a most weighty consideration, which is encountered in recent cosmological literature under the label of the anthropic principle.[24] The universe is indeed anthropocentric in a far deeper sense than the one which was discredited by the Copernican revolution.

Third, the cosmic background radiation strongly suggests, by eliminating the steady-state theory, that physical processes, even on the all-encompassing cosmic level, represent an unidirectional once-and-for-all phenomenon. Since the universe contains in all likelihood a finite amount of matter, and therefore its energy content too must be finite, such a process cannot be of infinite duration.[25] To be sure, it can have a past much longer than 20 billion years, the 'age' heedlessly assigned nowadays to the universe. Prior to the 'first three minutes', a charming misnomer, there could have been still unexplored long-term processes. One such was envisioned by Lemaître, who viewed the primeval atom as being in a kind of radioactive transformation for some 60 billion years before the actually observed expansion started.[26] Also, the study of elementary particle physics is far from being a closed book. During this century science has made two gigantic advances into the realm of matter and found that the realm of the atom, with dimensions a million times smaller than ordinary matter, was still a million times larger than the realm of the atomic nucleus. In the same way, science can still find a state of matter in which the fraction of a second can accommodate as many events as can thousands of years when it comes to facts of ordinary history.

The hope that the actually observed cosmic process (expansion) may repeat itself endlessly in the form of an oscillating universe, rests on grounds that are extremely weak from the scientific viewpoint. Observational evidence gives no encouragement to the expectation fondly entertained by some avowedly materialist cosmologists that the present expansion will turn into contraction. But even if this were to become plausible, there is at present no scientific possibility for the turning of that contraction into another expansion and for securing thereby a perennial oscillation for the universe.[27] At the end of that contraction all matter would rush together so as to produce a gigantic black hole. While its immense gravitational hold would foreclose another expansion, particular black holes, into which stars and galaxies aggregate as the present expansion goes on, are destined to dissipate themselves into sheer radiation, once the expansion reaches an effective temperature well-nigh absolute zero. Physics at present provides no glimpse into what may take place in a universe turned into radiation at a still unfathomable future.

While it is utterly futile to speculate about the time allotted to the universe, its state at any time shows characteristics which make it a beauty *par excellence*, a cosmos in short. It has supreme coherence from the very small to the very large. It is a consistent unity free of debilitating paradoxes. It is beautifully proportioned into layers or dimensions and yet all of them are in perfect interaction. It is on an evolutionary track which is firmly set and is heading in a specific direction. The evolution of the universe does not, contrary to the cliché phrase, have its origin in a chaos, that is, in an entity standing for confusion, disorder, and complete absence of specificity. Such an entity is unthinkable in the strict sense. Complete chaos, universal chance, and the like are contradictory notions. There is no such thing as absolute randomness. Any probability theory must assume the existence of some basic regularity and specificity.

This is certainly true of any 'chaos' that has ever served as the starting point of a cosmogonical theory of any repute. Indeed, the chaos which was the starting point of Descartes' cosmogony had such specific characteristics as the original division (by the Creator) of the homogeneous primordial matter into domains corresponding in size to the average distance among stars. As to that matter, it quickly turned (again through some covert disposition by the Creator) into three kinds so as to secure the formation of stars with planets. The chaos which was the starting point of Kant's cosmogony was, contrary to almost all accounts of it, an even more specifically tailored entity. It was another matter that, to recall an almost century-old very apt remark, in Kant's cosmogony the 'primitive reign of chaos was little likely to terminate'.[28]

It was again another matter, though in a different sense, when Laplace made popular the view that the nebular fluid as observed by Herschel was the initial stage of cosmic evolution. Since Herschel's observations of nebulae gave a magnificent view of the various stages of their evolution, the fluid in question could indeed serve as a hypothetical starting point. Unfortunately, nothing was known, until the advent of spectroscopy a hundred years later, about the constitution of that fluid. But because it was in all appearance perfectly homogeneous throughout, the misconception could arise that the enormous specificity of the universe could issue from perfect homogeneity. It was that perfect homogeneity which in its allegedly infinite extension was readily taken by many for that kind of necessary existence which was in no need whatever of a Creator. The conclusion was false not only philosophically but also scientifically. Philosophically, it is still to be shown that undifferentiated sameness assures existence for matter. Scientifically, no entity is subject to experimentation and observation unless it displays inhomogeneities and changes, that is, happenings.

Laplace also made the celebrated remark that once the conditions of the universe at a given moment are fully known, all future states of it can be safely predicted.[29] He hardly suspected how right he was. Modern scientific cosmology possesses such a grasp of a singular coherent universe as to connect with astonishing skill and exactness its configurations billions of years apart and to infer its large-scale features from the properties of its smallest constituents with breath-taking precision. Thus from the values of the charge of the electron, of the mass of the proton, of Planck's constant, and of the speed of light, one can infer why the sky is blue, what are the limits of the size of stars, what is the maximum height of a mountain and the like. But these values, or basic constants, must be taken as given in at least a provisional manner. Beneath them science will possibly find other, more basic, constants which will startle the sensitive mind even more with their specificity and with their even wider explanatory power. This will bring further witness to the specificity of the totality of consistently interacting things, the universe. Whatever exists apart from that totality is of sheer irrelevance for science, precisely because of the lack of interaction. Speculations about the existence of worlds which do not interact, however slightly, with the existing world, which is known to exist in full coherence, should remain the privilege of those addicted to idle imageries.[30]

Beauty tempting

In facing up to such a universe – thoroughly coherent, strikingly specific in space and in time, in its entirety and in its details, and exclusive in its oneness – the most reasonable attitude seems to accept it as something given, and given by a Creator. But precisely because that cosmic givenness conjures up the spectre of true creation as its only explanation, there has never been a lack of response to the age-old temptation to show that the main features of the universe necessarily are what they are. It is this *a priori* necessity which has served for many a thinker and scientist as a foil against Creator and creation. The foil has always been very transparent and should, as will be seen, appear especially so in this age of science. Eddington's failure to derive on *a priori* grounds the total number of atoms in the universe in his *Fundamental Theory*, a brilliant *tour de force* of mathematical physics,[31] did not pour cold water on the sanguine hopes of others. Hardly a decade after Eddington's death in 1946, Oppenheimer admitted to having been led by the desire of knowing why the elementary particles were what they were and why there ought not to be others except the ones already known or the ones predicted on their basis.[32] Oppenheimer also suggested that such was the fondest motivation of not a few of his colleagues grappling with the basic questions of physics. A dozen or so years later Professor Weinberg stated:

> 'Different physicists have different motivations, and I can only speak with certainty about my own. To me the reason for spending so much effort and money on elementary particle research is not that particles are so interesting in themselves – if I wanted a perfect image of tedium, one million bubble chamber photographs would do very well – but rather that as far as we can tell, it is in the area of elementary particles and fields (and perhaps also of

> cosmology) that we will find the ultimate laws of nature, the few simple general principles which determine why all of nature is the way it is'.

The enterprise appeared to be hopeful to him because a possible unification of relativity and quantum mechanics could eliminate a large number of possible avenues: 'Nature somehow manages to be both relativistic and quantum mechanical: but those two requirements restrict it so much that it has only a limited choice of how to be – hopefully a very limited choice'.[33]

Hope was verging into assurance as Professor Murray Gell-Mann outlined his theory of quarks on October 7, 1976, at the Twelfth Nobel Conference, which I attended as a co-panelist, to an audience of several thousand to whom he held high the vista of a final theory of elementary particles to be formulated in the near future, perhaps within a few months. The theory, he argued, would be final in the sense that it would contain the (*a priori*) reasons why the set of elementary particles, and therefore the entire structure of the universe, can only be what it is.[34] Very recently, Professor Steven Hawking, considered by some of his peers as the most incisive physicist since Einstein, gave an interview a part of which is reported here with some misgivings in view of the journalistic inaccuracy which seems to increase in direct proportion to the number of tape-recorders available.

> 'My goal is obvious: complete understanding of the universe. But I think it would be very presumptuous to say that I expect to get the complete solution. One goal of my work has been to understand whether the universe has a meaning, and what our role is in it. I've always wanted to know why the universe exists at all, and what was there before the beginning. There may be ultimate answers, but if there are, I would be sorry if we were to find them. For

> my *own* sake I would like very much to find them, but their discovery would leave nothing for those coming after me to seek. Each generation builds on the advances of the previous generation, and this is as it should be. As human beings, we need the quest'.[35]

The wish to understand the universe completely can have a perfectly sound meaning from the scientific viewpoint. Insofar as mathematical physics aims at a full description of the quantitative aspects of the interaction of things, a complete accomplishment of that task is not intrinsically impossible. Insofar as the universe is a totality of consistently interacting things, or to use the words of the Book of Wisdom, a totality of things disposed according to measure, number, and weight,[36] there is nothing unfathomably mysterious about the universe. The universe needed by science is done a great injustice when it is mystified by scientists who confuse questions of mathematical physics with questions of metaphysics as if the tools of the former were appropriate for the latter. Professor Hawking's statement is a perfect illustration of that confusion. Is it not the old fallacy of shifting terrains illogically (*metabasis eis allon genos* – the Greeks of old would say), to imply that a theoretical physicist as such can hope to answer the question why the universe exists at all, to say nothing of trying to answer with the tools of mathematical physics or with any conceivable tool the question: what was there before the beginning?

This last question admits for an answer only a reference to the usefulness of introductory courses in logic where one deals with the propriety of defining one's terms. Taken in a strict sense, the term 'beginning' pre-empts any inquiry of its meaning as to what happened before. A passage from Saint Augustine's *Confessions* is particularly relevant in this connection both because of its dignified reasonability and also because it is often abused by cosmologists who know it only second-hand:

> 'My answer to those who ask "What was God doing before he created heaven and earth?" is not "He was preparing hell for people who pry into mysteries." This frivolous retort has been made before now, so we are told, in order to evade the point of the question. But it is one thing to make fun of the questioner and another to find the answer. So I shall refrain from giving this reply. For in matters of which I am ignorant I would rather admit the fact than gain credit by giving the wrong answer and making a laughingstock of a man who asks a serious question'.[37]

Beauty vindicated

While ignorance in a matter that cannot be known is a normal state of affairs, the case is different with ignorance in matters that can be known without much effort and with full conclusiveness. A broad awareness among cosmologists is still to arise concerning the very real possibility of knowing conclusively the intrinsic impossibility of knowing that the universe can only be what it is. The latter knowledge implies the necessary and exclusive character of the existing universe and can have therefore its source only in *a priori* considerations. To know the universe in such a way is not only very different from knowing all its quantitative correlations (a project not intrinsically impossible, as was hinted above), but also rigorously impossible. The ambition to formulate an explanatory framefork in which the world appears to be a necessary form of existence is not new. It is as old as physics itself. The two major types of physics, the Aristotelian and the Newtonian, which preceded modern physics, were based respectively on the conviction that the notions of organism and mechanism were basic, necessary forms of intelligibility to which reality had to conform.[38] Organismic physics and mechanistic physics are things of the past that need not detain us. Modern physics is not only very much alive but also incomparably superior in its attainments to its predecessors. Can this physics give us an understanding about the cosmos which is *a priori*, that is, the necessary form of understanding and by inference of physical existence? The answer to this question can only be negative as long as Gödel's incompleteness theorem is true. According to that theorem – its truth has withstood much intensive probing since its formulation in 1930 – no sufficiently broad (non-trivial) set of arithmetical propositions can have its proof of consistency within itself. Clearly then no scientific cosmology, which of necessity must be highly mathematical, can have its proof of consistency within itself as far as mathematics goes. In the absence of such consistency, all cosmological models, all theories of elementary particles, including the theory of quarks and gluons which received a remarkable experimental verification recently,[39] fall inherently short of being that theory which shows in virtue of its *a priori* truth that the world can only be what it is and nothing else. This is true even if the theory happened to account with perfect accuracy for all phenomena of the physical world known at a particular time. There is no theoretical necessity why no new phenomena would in the future come to man's notice. The world of phenomena and its beauty will appear completely exhausted only to those who at one point or another declare that further surprises are *a priori* impossible as far as observation and experiment are concerned.

Tellingly enough, this broadest and deepest philosophical significance of Gödel's theorem remained unnoticed for a long time. Even a Carnap, who urged, though in vain,[40] his colleagues in the Vienna Circle to study the theorem of Gödel, himself a member of the Circle, failed to realize that it spelt the death-knell on logical positivism and especially on his own confident *a priori* account of the universe set forth in his first major work, published in 1928, on the 'logical build-up of the universe'.[41] As one would expect, logical positivists did their best to take the sting out of Gödel's theorem when from the 1950s on its fundamental importance began to be generally recognized. Such an approach could not remain immune from contradictions and oversights. Glaring examples of both can be found in the most widely read exposition of Gödel's theorem, written by Nagel and Newman.[42] They recognized on the one hand that 'there are innumerable problems in elementary number theory' that cannot be answered by computers 'however intricate and ingenious their built-in mechanism may be and however rapid their operations'. On the other hand they also claimed that the replacement of human minds by robots was not intrinsically impossible, but merely that there was no 'immediate prospect' for this. These two claims, contradictory enough, were followed by a claim which was contradiction in itself: 'Nor do the *inherent* limitations of calculating machines imply that we cannot hope to explain living matter and human reason in physical and chemical terms' (italics added). As to the oversights, one is implied in their statement: Gödel's theorem means 'that the resources of the human intellect have not been, and cannot be, fully formalized, and that new principles of demonstration forever await invention and discovery'. Since this is true, the inference is obvious, though it wholly escaped Nagel and Newman, that the effort of the cosmologist, who wants to demonstrate on *a priori* grounds (that is, through the full formalization of thought) why the universe is what it is and cannot be anything else, is doomed to failure.

This correlation of cosmology and Gödel's theorem, a feat hardly more difficult than putting two and two together, has been widely available in print in a book[43] which soon after its publication was recommended in a prominent context by Professor W. Heitler of quantum mechanics fame, as a 'compulsory reading for all scientists, students, and professors'.[44] If silence about its pages containing that correlation is an indication, it was not read by those who were most in need of it. Whether my other discussions of Gödel's theorem in the same vein, and in no less widely available contexts, will stir comments, is rather doubtful. The climate of thought in our times is not at all favourable for a recognition of reason's ability to bring within sight the contingency of the universe and its *raison d'être*, its having been created by a Being truly necessary. This climate of thought has for its prisoners even many of those who call themselves theists nowadays.

Still, if there was an age which stood in need of awareness of Gödel's theorem it is this age of ours in which the fashionable form of philosophizing hardly reaches beyond the confines of exercises in mere logic. Logic, it is well to remember, is impotent to put one in contact with reality, let alone with the reality of the cosmos and its supremely rational beauty. While the barren futility of any *a priori* speculation about the cosmos is powerfully revealed through Gödel's theorem, awareness of it is not necessary for grasping the givenness of the cosmos. A modest degree of intellectual humility with respect to facts is enough. A case in point is Einstein. Although he had on occasion been in the mood of pontificating about the cosmos as if he had been its maker, he remained fully aware of the fact that the ultimate truth of theories lies not in their preconceived beauty but in their agreement with the facts of observation. Without the support of facts any theory, as Einstein put it in 1920 about his General Relativity, would turn 'into mere dust and ashes'.[45] Trusting as he was in the power of the mind to establish immensely effective theories as to the pattern into which the facts of the universe would fit, he never went as far as to hint that the mind created those facts and their totality, the universe. He certainly never claimed that the rational character, or the consistency and coherence of the facts of the universe, was a creation of the mind and not a characteristic of a reality existing independently of the mind. He also knew about the ultimate thrust of this position of his. In his always straightforward though never trite diction he spoke of it in his letter of January 1, 1951 to Maurice Solovine, an old friend, who expressed concern about Einstein's possible drifting from positivism toward metaphysics and theism. Einstein reassured him but in words which only proved once more that protestations spoke more loudly than plain avowals. Clearly, not even an Einstein could be an exception to the power of logic inherent in an outlook on nature which proved itself exceptionally fruitful from the scientific viewpoint: 'I have never found a better expression than the expression "religious" for this trust in the rational nature of reality and of its peculiar accessibility to the human mind. Where this trust is lacking science degenerates into an uninspired procedure. Let the devil care if the priests make capital out of this. There is no remedy for that'.[46] Could there be a more telling recognition that nothing could remedy the predicament in which saying cosmos truthfully had to be followed by saying Creator as well?

The force of logic pushing Einstein toward theism made Solovine apprehensive, but he should have started worrying twenty or thirty years earlier. From the 1920s on Einstein repeatedly made declarations that were so many signals of his escape from the clutches of Mach's positivism, but Solovine failed to notice them. In 1952 Einstein wrote him another letter which, contrary to its intention, must have made all too clear to Solovine the full force of logic which urges a most creative scientist to say Creator after he had confessed his faith in the existence of the cosmos. Indeed, Einstein could not have been more explicit in recognizing the force of that urge. Of course, he put up a brave resistance, perhaps to reassure his worrying friend:

> 'You find it surprising that I think of the comprehensibility of the world (in so far as we are entitled to speak of such world) as a miracle or an eternal mystery. But surely, *a priori*, one should expect the world to be chaotic, not to be grasped by thought in any way. One might (indeed one *should*) expect that the world evidence itself as lawful only so far as we grasp it in an orderly fashion. This would be a sort of order like the alphabetical order of words of a language. On the other hand, the kind of order created, for example, by Newton's gravitational theory is of a very different character. Even if the axioms of the theory are posited by man, the success of such a procedure supposes in the objective world a high degree of order which we are in no way entitled to expect *a priori*. Therein lies the "miracle" which becomes more and more evident as our knowledge develops . . . And here is the weak point of positivists and of professional atheists, who feel happy because they think that they have not only pre-empted the world of the divine, but also of the miraculous. Curiously, we have to be resigned to recognizing the "miracle" without having any legitimate way of getting any further. I have to add the last point explicitly, lest you think that, weakened by age, I have fallen into the hands of priests'.[47]

The merits of Einstein's claim, that there is no way of going beyond the cosmos to its Creator, a rehash of a worn-out philosophical cliché, will be discussed in another chapter. In this chapter, which deals with the beauty of the cosmos as revealed through science, it should suffice to recapitulate briefly the chief characteristics of the universe which make scientific work possible. First, the material entities observed by science must be real, that is, existing independently of the observer. Were not such the case, each observer would create his own facts, a result banishing each observer to solipsism, the strictest solitary confinement imaginable. No observer reduced to that confinement can lay a claim to an exchange of his views with other observers, who, at best, are, together with their worlds, the creation of his own mind. Second, the material entities must have a coherent rationality. They must be governed by laws which can be formulated in a quantitative framework, and they must have a validity which transcends the limits of any particular time and location. Third, those entities, because they are governed by consistent laws, must form a coherent whole, that is, must be subject to a consistent interaction. The existence of any material entity which does not interact in a coherent way with the known world is utterly irrelevant for science. For science there is only one universe. Science has no room for island universes and multiple worlds if these stand outside the realm of physical interaction, be that interaction gravitational, electromagnetic, or of any conceivable kind. Fourth, the form in which that coherent wholeness, or universe, does exist, cannot be considered, partly in view of Gödel's theorem, a necessary form of existence. It is only one among countless others that are conceivable. As to the question why such a universe does in fact exist, science has no answer. It cannot even answer the far less deep question whether the duration of that world is infinite or not.

These four features of the universe are indispensable not only for making the notion of the universe worthy of its etymology, the converging of many into unity, but also for making science possible. Neither is science conceivable without any of them, nor are they conceivable without one another as long as one aims at a rational discourse about the universe. Those four features form a single basic proposition which must be assented to unconditionally if any further proposition, that is, a message addressed from one human being to another, is to make sense. A proposition which demands unconditional assent has since long been denoted as a dogma. That basic proposition certainly functions as an initial dogma in that superbly articulated creed about the reality and rationality of the universe which is science. Not surprisingly, the ultimate justification of that dogma can be found, both historically and philosophically, only in that article of faith, the dogma of Creation, which is the basis of all genuine and reasonable dogmatic propositions.

CHAPTER THREE

THE DOGMA OF CREATION

Beginnings and beginning

The dogma of creation is the first article of the Creed and also its most neglected article. Such a state of affairs will appear both anomalous and revealing to anyone ready to ponder a few truisms. If there is any truth in the poignant motto, 'In my end is my beginning',[1] it is only so because the beginning always anticipates the end. What is true of initial conditions in a physical process is true also about the very first statement of any thinker determined not to talk from both corners of his mouth. The inner force of logic comes into play as soon as the starting point of a discourse has been enunciated. There is indeed a felicitous touch in the expression, 'train of thought'. It does justice not so much to one-track minds caught in the triviality of logic chopping, as to all those whose discourse is directed to a specific destination along carefully laid down lines.

This certainly must be true of any serious philosopher but also of any prophet. For if a prophet is not a false one, his message must be about truth, and although truth is always more than logic, logic is an ingredient of any truth. This is why the first article of the Creed – I believe in God, the Father Almighty, Maker of Heaven and Earth – makes the rest possible. Without Creation, and a Creation by God who is Father, there is no possibility of a discourse about Incarnation, Redemption, and final consummation in a New Heaven and Earth, the great prophecy of the Creed. Conversely, no attack on any article of the Creed has ever been made without some consequence for the meaning of its first article. And when an article of faith, such as Redemption or Incarnation, is simply denied, the ultimate result, more often than not, is a denial of the proposition that whatever exists does so because it is part of a totality of things which the Creator brought out of nothing into existence. This is, of course, only the reverse of a pattern verified time and again, namely, that one's notion of 'in the beginning' sets the pattern of one's thinking about any beginning.

Yahweh the Lord

An espousal of the first article of the Creed sets one's thinking on a very particular, nay unique, course also because the recognition of one's and of all things' createdness is the sole alternative to the assumption that the existence of things is self-explaining. Not surprisingly, the espousal of the first article of the Creed appears as a very unique chain of events across history. The first link of that chain is almost impossible to specify, although the chain itself has a uniqueness which thrusts itself into bold relief by its dramatic character. No drama is more poignant than a mother's defiance of the death of her seven sons and finally her own. Its written record is in the Second Book of Maccabees, composed in the wake of the uprising which the Maccabee brothers led against Hellenization in the middle of the second century BC. The drama reaches its height, both emotionally and conceptually, when the youngest of the seven is to face the choice between clemency and execution. Contemptuous of the king's suggestion to induce even her youngest son into apostasy, the mother, who had already witnessed and encouraged the martyrdom of the six elder ones, bends over the youngest with the exhortation: 'I implore you, my child, observe heaven and earth, consider all that is in them, and acknowledge that God made them out of what did not exist, and that mankind comes into being in the same way'.[2]

The declaration that heaven and earth and all that is in them were made by God 'out of what did not exist' is much too clear to permit any doubt that one is faced here with the notion of *creatio ex nihilo.* The matter-of-fact appearance of this notion, as speculative as a notion can be, in a distinctly non-speculative context is equally significant.[3] The spontaneity of the author of the Second Book of Maccabees, in all likelihood a member of the very large Jewish community in Alexandria, seems to reflect an already widely shared conviction that the almost abstract way of contrasting that which exists with that which does not exist was a very appropriate way of expressing the full Lordship of Yahweh over all things, a Lordship which had already been voiced in many concrete forms throughout the Old Testament. All those forms, mostly picturesque metaphors or plain enumerations of the main classes of very concrete things, served to emphasize that all things were made by Yahweh in such a way as to owe to him their very existence and being. Such was the way of the Hebrews of old to convey an idea whose full meaning was later unfolded through the expression *creatio ex nihilo.*

The complete, unrestricted sovereignty of Yahweh over heaven and earth, an idiomatic Hebrew equivalent of the relatively abstract 'all' or 'everything', was an axiomatic article of faith for the Hebrews, or rather for their stubbornly persisting minority which did not succumb to the ever present temptation of nature worship or idolatry. The notion of that sovereignty could not be borrowed by them from their Semitic kin and certainly not from that famed Babylonian story of creation, the *Enuma Elish.*[4] In that story Marduk, who forms with his word the present shape of the world, is a Johnny-come-lately. He is the offshoot of three generations of gods, the first of which is the couple, Apsu and Tiamat, representing the heaven (generating power) and the earth (universal motherhood), respectively. Either singly or together, they are not the lord of all. They merely represent the hypothetical first stage of all, a stage which is not first in a strict sense. It is merely, as the story itself suggests, especially when taken together with the symbolism of the yearly Akitu festival in Babylon, a relatively first stage in a perennially recurring state of affairs. Being cyclic, in such a process no stage is strictly first or primordial. The process is a transition from a quiet inchoate phase through a violently chaotic one to an orderly stage which is bound to collapse and transform itself back into the quiet inchoate phase. The violently chaotic stage is, tellingly enough, the most prominent part of the *Enumah Elish.*[5] It is told in terms of a death-to-life struggle within the family of gods. The high point of the story is the savage dismemberment of Tiamat's body which provides the material for the various parts of the actual world. The shaping of that world is done at the command of Marduk who emerges as the strong man among the infighting gods and godesses. Unlike Yahweh, Marduk is not an absolute Lord, a Lord utterly superior to everything visible and invisible.

To be sure, the name Yahweh was not born on the lips of Abraham. But even according to the oldest parts of the Old Testament, the chapters of Genesis recounting the lives of the patriarchs, the Lord is never a mere *primus inter pares,* or the mere top of a divine hierarchy, let alone one of two supreme powers, or principles, caught in an eternal struggle. His is rather an unrestricted sovereignty over all that exists and happens. His most specific name by which he enters into a covenant with Abraham's progeny, a covenant to give ultimate meaning to all that happens, in human as well as in cosmic history, is YAHWEH (HE WHO IS or I AM WHO

AM).[6] Such a name, which is undoubtedly most unexpected, is far more powerful to convey the primordial importance of existence than even such modestly abstract philosophical expression as 'existence itself', applicable though it be to God alone. In fact, it is rather fortunate that not even that much trace of philosophy is found in the biblical doctrine of God and especially in that chapter of Exodus in which God's specific name is given. The presence of any expression harking back to 'professional philosophy' would easily imply that the 'metaphysics of Exodus'[7] is no less human than all metaphysical systems in which there is no room for it. The stark concreteness of the name Yahweh eliminates such a suspicion. Such a name is too unique and unexpected to justify efforts aimed at tracing it back to cultural or sociological contexts within which even the most unexpected must appear inevitable.

Yet, defying as it does cultural and sociological categories and understanding, the name Yahweh is perfectly understandable once it is given from above. In a sense the process is like the first circumnavigation of the globe. Once it is done, all lands and seas fall into their proper places and all alternatives, such as that of a flat earth, reveal their utter inadequacy. But unlike Magellan's trip, the name Yahweh, which represents the sudden rise to the highest peak of metaphysics, is not the fruit of planning, of lengthy preparations. It is certainly not a borrowing from Egyptian lore which reached its peak, very low in comparison, when a mere visible thing, the sun, was under Akhenaton's brief reign the object of exclusive worship, a very pale image of monotheism. The name Yahweh is not intimated in advance in the narrative of which it is a pivotal and integral part. Equally important, in parts that were added in a more reflective manner to the original narrative, there are no attempts whatever to pour a theology around the name

Yahweh. Undoubtedly, the temptation of doing so must have been enormous. Human nature has its built-in temptations, of which the apparently noblest incubates in theologians. Our times gave renewed and vast evidence of their proneness to turn theology into a reduction of toweringly unique, stunningly unexpected 'un-natural' single facts to generalities all too readily expected by human nature.

Such a theology, or rather theological disease, was at work, as will be seen, in the case of Philo, the great first-century Jewish scholar in Alexandria. But he was clearly his own voice and not that of a tradition which remained faithful to type, a faithfulness of which the books of the Old Testament are the written record. In that record there was room for the words uttered by the mother of those heroic brothers, because those words are a simple affirmation of concrete truth: He made all out of what did not exist. Even in connection with the notion of logos which was incomparably better developed among the Greeks than the notion of existence, the biblical assimilation is an example of judicious restraint. Much the same can be observed in those parts of Genesis which were added to the narrative parts in post-Exilic times, but still prior to the advent of Hellenism following the death of Alexander the Great.

Those additions bear the impact of a direct exposure during the captivity in Babylon to a culture which, though Semitic, was with respect to crafts and organization on a level much higher than what had ever been achieved in Judea. Tellingly, the most pervasive and characteristic features of Babylonian religion, such as necromancy and astrology, failed to break the continuity of an already long religious tradition whose pivotal tenet was Yahweh's supreme and exclusive sovereignty over all. The tradition was not simply equivalent to the Jewish people. Even according to biblical data, a considerable portion of the people had always been less than committed to Yahweh and true worship. The agnosticism and materialism of so many Jews in modern times had long been anticipated by their forebears' worship of the golden calf, by their cavorting with the idols of the Canaanites, and by their surrender to the ideals of Hellenism. Those who speak of a Hebrew religious genius pay homage to the hollow slogans of sociology, psychology, and genetics, but not to the facts of ancient Jewish society and history.[8] Instead of generating a unique belief, that society was sustained by a faith which was articulated and reaffirmed through the unexpected rise of an outstanding religious teacher, usually a prophet.

Genesis revisited

The impact of such a person is clearly visible in the redaction of the first chapter of Genesis. That chapter, whose renown is in part due to its uninformed abuse and ridicule by those who more often than not only reveal their lack of elementary information on the subject. The chapter, composed in post-Exilic times, is remarkably untouched by the confusions teeming in the *Enumah Elish.* In the first chapter of Genesis there are no infighting gods, no gory dismemberment of a divine mother, no threat of collapse. The inchoate or chaos (*tohu vabohu*) is retained, but not as a symbol of divine principle, let alone of an evil principle on equal footing with what is good. The chaos in Genesis chapter one is wholly subject to God, whose spirit hovers over it, and so are subject to him all general and particular features of the world which are called forth out of the chaos by his sovereign command. All that is produced by God is declared emphatically to be good.

Genesis chapter one does not say that the chaos itself was produced out of nothing. It is not even stated there that the chaos was made by God in any specific sense. Only the rest of creation is made by God and through an action which is conveyed by the word *bārā'*. Although etymologically its meaning is to carve, hardly a divine prerogative, it is used throughout the Old Testament to denote an action proper to the sovereign Lord of all.[9] Such an action reveals its deepest aspect in the making of something out of what does not exist, but this development in semantics came only under the impact of the confrontation with Hellenism and its philosophy in which the world was eternal and divine. The development was, as will be seen, thoroughly biblical and wholly logical. The full sovereignty of the Lord over heaven and earth, that is, over everything, as emphasized in Genesis chapter one, is void of meaning if anything can possess the slightest measure of independence of him. On the contrary, everything is fully dependent on him and the dictum *'creatio ex nihilo'* will merely reveal the very fullness of that dependence.

The full sovereignty of the Lord over everything is set forth in Genesis chapter one in a manner which is the strongest hint that once again more than mere human insight was at work. Yet it is that manner which is the source of the most persistent misunderstanding of what the first chapter of Genesis is about. The manner is primitive simplicity. The primitiveness of the world-view of Genesis chapter one has often been held up for ridicule,[10] a rather grotesque pastime because it merely reveals the proverbial shallowness of hindsight, if not plain ignorance. The first chapter of Genesis, this classic statement of the dogma of Creation, is no match for the sophistication (though very nebulous) of the nebular hypothesis, and not even for the cosmogonical hunches of the Ionians of old still caught in geocentrism. The picture of the world as a house or rather a vast tent erected on flat land floating on abysmal waters, is certainly primitive, though very genuine as far as time and place are concerned. No less primitive is the presentation of the operation of the Lord of all as being analogous to the procedure followed by any ordinary workman who in the manner of all Semites works for six days and rests on the seventh. Again, what is more primitive (though extremely logical) than to have light first and then turn the piled-up building material, still undifferentiated, into a structure? The world-structure given in Genesis chapter one as the work of the second and third days is indeed so primitively told as to leave out the gigantic columns which, according to other parts of the Bible and in agreement with the Babylonian world-view, rise at the two ends of the dry land to support the heavenly vault. The work of the fourth, fifth and sixth days is taken up by the ornamentation of that structure and in a way no less primitive than the erection of the structure itself and its main purpose: the separation of the upper waters from the lower waters. The sun, moon, and the stars are the décor of the vault, whereas the birds, fish, and animals are the décor and furnishing of the lower part of the world-tent which is then ready to receive man, its divinely appointed master. All this is told in a sweeping cadence, in the form of a thematic poem. All its strophes are introduced with a reference to God's almighty decision, 'Let there be . . .', and concluded with his supreme approval: 'And God saw that it was good.'

Such is the message of the first chapter of Genesis, that often misunderstood classic statement of the dogma of creation. What certainly cannot be misunderstood about it is its place of honour in the entire Scriptural corpus. Soon after its composition it was put at the head of an already centuries-old and very anthropomorphic creation story (Genesis 2 and 3) so as to warn about a divine factor operative beneath very human forms. It also stands in front of all books of biblical revelation as a warning that the salvation to be achieved through revelation rests on the dogma of creation. As expressed in the first chapter of Genesis that dogma should seem stamped with utmost lucidity to any modern possessed of that minimum of enlightenment which makes one realize that a Jewish teacher of the fifth century BC would have been the victim of falling back on a *Deus ex machina* had he spoken of creation in a more theological or philosophical way. He spoke of it in the only way natural to him, a way in one crucial aspect also exclusively natural to all his contemporary world. For with Semites, as well as with Greeks and others, it was an axiom that the shapely world was preceded by an unshapely one, or chaos. Although very important philosophical questions could be raised about this precedence, it was not natural for a Jewish teacher at that time to raise them. By speaking of the origin of things in his natural way, the author of Genesis chapter one was therefore able to produce a text which could be an instrument of a genuinely supernatural intervention or revelation. About such revelation the most divine and most human is embodied in the dictum that the grace of God does not simply overpower and thereby destroy human nature but elevates it and especially its most noble part, the mind.

The coming of Jesus

The way in which revelation elevates or instructs the mind about creation has in it that subtle, almost intangible character which is the very core of all great dramas. There the source of action is never complete virtue or utter vice, but a subtle insufficiency in the hero's moral character. In the same way, revelation does not provide truth according to the categories of a philosophy which does not perceive the mirage of tautology in the apparently full light of logical clarity. Revelation does not overpower man's mind by turning it into an automaton or by lifting it from its natural condition which is to be puzzled by truth. Indeed when revelation reached its peak with the coming of the Master of Nazareth, the puzzlement was deeper than ever. Could it really be true that anyone who saw Jesus saw the Father? Undoubtedly, Jesus' central theme was the Father in heaven, not a new theme for a pious Jew.[11] But unlike in the Old Testament, the Father, whom Jesus addressed in one of his few recorded prayerful utterances as the Lord of heaven and earth,[12] was depicted by him with an immediacy never witnessed before. In Jesus' preaching, the very possibility and actuality of redemption were justified by his incomparable insistence that God, the Lord of all, the Creator, was a Father in the most incomparable and touching sense of that word.

The touch was that of mercy, a word often confused with condescendent commiseration or with a contemptuous dismissal of charge. The divine mercy of which Jesus spoke was not tainted by such all too human traits. This is why it could appeal so forcefully to mere humans who saw it displayed not only in inimitable words but also in a life to be imitated as closely as possible. The purpose of that life was to bring to its fullness the vitality inherent in the dogma of creation, a vitality which derived from the fact that from the very start of the Old Testament its references to creation by God were so many recognitions of divine providence as well. The identity of these two themes had already received its thematic expression in the Book of Wisdom, usually remembered for a famous passage which reproaches pagan students of nature for their failure to recognize the Maker of All from the beauty and regularity of nature, especially evident in the realm of stars.[13] The passage, a shorthand for natural theology, is not, however, given justice when separated from its context which is about God's providence as revealed especially through the exodus from Egypt of the people of the Covenant.

The context is no accident. It is rather of a piece with an already long tradition in which the union of cosmic perspective and the perspective of salvation implies the pedagogical primacy of the latter over the former. The primacy rests on the fact that the details of salvation history cut much more concretely into man's flesh and blood than do details of cosmic history however picturesquely told. The same primacy is very evident in Jesus' words and deeds which therefore become the fulfilment of the Covenant also by bringing forth in a personal concreteness the vitality of which God's creation was susceptible. Such was in fact the highest vision which Jesus' disciples could form of him. In Paul's words, Christ, the Saviour, is the very principle of creation through whom everything comes into being.[14] According to John nothing that was made came into being without the Word (Logos) which is Christ.[15] But in Christ, who is also the principle of new spiritual creation, all beings have a higher destiny which is expressed by their trend toward him.

This centrality of Christ, who rules over the entire cosmos destined for salvation, gives to the dogma of creation a new vitality. Its measure can be grasped by the fact that unlike the Old Testament, the New Covenant made a universal impact. The confrontation between Hellenism and the Old Testament resembled the resistance of a small island to mighty waters in whose advance it appeared as a mere incident not worthy of attention. Indeed, just as there is no trace of a Babylonian refutation of the Jewish Creed resting on the exclusive sovereignty of Yahweh, Hellenistic philosophers felt no need to justify their basic tenets with an eye on that sovereignty. When around the middle of the second century AD, Celsus wrote his famous defense of polytheism,[16] he presented as its real threat not the Jews, already for five centuries in Alexandria and in impressive numbers, but the Christians, not yet a century there and far less numerous. To be sure, Celsus is fully aware of the Christians' indebtedness to the Jews concerning monotheism. But in a crucial sense the Jews are not, according to Celsus, different from other nations who stick to their religious traditions and let others live according to their own, a pattern, which for Celsus reveals the variety in which a divine Nature should evidence itself. The conflict of Christian faith and patterns with that divine Nature becomes most evident through the Christians' belief in Jesus as the Incarnate God who spoke to men to the extent of coming down to the earth. Such a notion of revelation is far more vivid and seductive than what is offered in Jewish monotheism and is therefore a far greater threat to that Nature which, being eternal and divine, bars the possibility of revelation.

The doctrine of Incarnation throws into powerful relief not only the dogma of creation, but also the dogma of a special course of Providence across history. The latter was equally inadmissible within the world-view advocated by Celsus. There the ultimate principle of the universe is its intrinsic eternity, which is, of course, radically contrary to the dogma of creation. No wonder that the other major refutation of Christianity from Hellenistic times had for its main topic the eternity of the world. Written by Proclus, the last great figure of Hellenistic thought, about three hundred years after Celsus, it received a lengthy reply from Philoponus, one of the last teachers at the Academy in Athens, who became a Christian around 529. The oblivion into which Philoponus's works have fallen is deplorable as they contain not a few incisive remarks which show how the doctrine of creation could inspire scientifically fruitful reflections about the physical world. That doctrine made its heavy impact even in those writings of Philoponus which are void, as is the case with his reply to Proclus, of any reference to Christian creed or Bible. A sustained, though mostly abstract and at times abstruse argumentation against the eternity of the world, would have been inconceivable without that 'metaphysics of Exodus', which unmistakably lurks behind Philoponus's animated insistance, that God alone *is* and that '*was*' and '*will be*' can never be predicated of him.[17] At any rate, the historical victory of the dogma of creation over pagan doctrines stressing the eternity and, by implication, the uncreatedness of the world was not so much the fruit of philosophizing, however ably, but of the riveting of Christian consciousness on the uniqueness and cosmic relevance of Christ, the God Incarnate. In an eternal world, subject to an endless succession of great cycles or Great Years, the death of Christ could neither be unique nor relevant, a consequence revolting to any Christian. But the same consequence threatened, as Origen, Augustine, and other Church Fathers hastened to point out, the meaning of the death of Socrates, that most influential and enormously respected event of classical pagan antiquity.[18]

Creatio ex nihilo

Such was the dramatic way in which unconditional commitment to a unique, once-and-for-all salvation in Christ, that is, to the uniqueness of salvation history, became the shield protecting the uniqueness of the history of the cosmos and ultimately the doctrine of its creation out of nothing. The same happened along another line, less dramatic in a sense, but certainly more decisive. It is connected with the conceptual refinements of the doctrine of Incarnation around which raged the great inner debates of the early Church. A tracing out of that line, through which the doctrine of *creatio ex nihilo* became a hallowed tenet in Patristic literature, is also very instructive of the slow development in which a specific meaning is finally and firmly attached to a word not endowed with it initially. The word is 'to create', which after so many Christian centuries cannot help but evoke the meaning of an action whereby something is not merely transformed from one state to another, but is made to exist. The word 'to create' originally had, as a derivative of *crescere*, only the meaning of making something to grow. No more appropriate for denoting a creation out of nothing were other Latin and Greek words used to indicate the origin of the universe as stated in the Bible. The word 'to make', which either in its Greek (*poiein*) or in its Latin (*facere*) equivalent was heavily used by the Church Fathers, is particularly instructive in this respect. That *facere* does not exclude creation in a strict sense can easily be seen from the phrase, *Factorem coeli et terrae*, 'Maker of heaven and earth', the very opening of the Creed. Yet, in spite of this hallowed connotation, its primary meaning has remained what it has always been, a mere reshaping of something already in existence. As for *poiein*, its adaptability to express creation can be gathered from its derivative, poetry, but artistic creation falls very short of what is implied in creation properly so called. A third word, *genesis*, widely used in modern English, reveals much the same fluidity of meaning to anyone ready to look up its definition in any good dictionary.[19] In its original Greek form it had no less a wide variety of meanings. One of them had taken on a major significance ever since Plato spoke of the cosmos as something ungenerated or having had no genesis.[20] In doing so Plato merely echoed an already long tradition and put an authoritative stamp on it which was never seriously challenged through classical antiquity. Being ungenerated, the universe could not have a beginning and had therefore to be eternal.

A universe without a beginning (genesis) was a very different phrase from 'in the beginning', a phrase opening the Bible. 'In the beginning' strongly suggested that the universe had a beginning and that therefore Plato was wrong in asserting its unbegotten, ungenerated character. But no sooner had this inference been made by a Christian, than he was confronted by the hallowed phrase, 'the only begotten son of God', the very definition of Christ in John's Gospel. Clearly, a Christian was barred from denying the truth of the Platonic expression 'unbegotten cosmos' by turning it into its semantic opposite, a 'begotten cosmos', that is, a cosmos that had a beginning because it was generated, begotten, by God. It was impossible to load the same word with two meanings so different as coming into being out of nothing (the origin of the world) and the eternal birth of a Son consubstantial with the Father.

That the Christian notion of creation (not yet tied to the word *creare*) could not be grafted onto the words 'to generate' or 'to beget' was a felicitous outcome, because it helped to oppose various and very powerful heterodoxies such as Gnosticism and Arianism. The Gnostics invariably spoke of the origin of the world as an emanation from God and for them the expression 'to beget the world' was most appropriate. To emphasize the orthodox doctrine of the essential difference between God and world, the expression 'to beget the world', already abused in support of emanationism, had therefore to be abandoned. As for the word 'to make', it was clearly too loaded with connotations concerning purely human actions to be able to carry in itself a much superior meaning unless it was made part of an entire expression, such as 'Maker of heaven and earth'. But this expression, though hallowed, could not, precisely because of Genesis chapter one, directly assume the meaning that the undifferentiated material (chaos, *tohu vabohu*) also had a beginning.

The thread which helped Christian orthodoxy out of an etymological labyrinth was provided by the now all too well-known expression *ex nihilo* (out of nothing) which was attached to the words *facere* (to make), *condere* (to establish), and *creare* (to make grow). The first to put forward such a combination was Tertullian who also used the form *de nihilo.*[21] The combination's advantage lay chiefly in the graphic conciseness of the Latin idiom as compared with the less succinct but more precise Greek equivalents *ek tou me ontos* (from the non-existing) and *ex ouk ontōn* (from those not being) already in use among Greek Christian writers a generation or two before Tertullian. The chief of them was the author of the *Shepherd of Hermas* to which both Aristides and Saint Theophilos of Antioch were indebted. This certainly reveals something of the importance of the *Shepherd of Hermas,* which had indeed been held by some for one of the canonical writings composing the New Testament. A main reason for this was its echoing many parts of the New Testament and such rather recent books of the Old Testament as the Book of Wisdom and the two Books of Maccabees. There can be hardly any doubt that the declaration 'he made everything from what was not existing to exist' in the *Shepherd of Hermas* had for its source the words the mother of those seven brothers whose heroic story is given in the Second Book of Maccabees.

What is most telling about that declaration is its entry into Christian literature from the very beginning and the matter-of-fact manner in which that entry had been accomplished. There is no hesitation whatever on the part of those writers concerning the appropriateness of 'from what was not existing' to convey the true meaning of the making of the world by Almighty God. This is certainly a striking contrast with the Greeks' attitude toward the notion of creation out of nothing. That attitude was a spontaneous dismissal, nay scorn, of the whole idea. This is why it occurs only half a dozen times in the vast corpus of classical Greek literature. In Greek texts composed in pre-Christian times it occurs on three occasions. The latest of them, a remark of Plutarch, is the least ambiguous. In his commentary on the procreation of souls as dealt with in Plato's *Timaeus*, Plutarch leaves no room for doubt: 'the creation [of the world] did not take place out of the non-existent, but out of that which was not beautiful and sufficient' that is, from the chaos.[22] The absence of any reference to God should be of no surprise. The Greeks of old simply could not think of a God who had a truly creative power over the universe. More often than not God, or rather the divine, was merely the noblest part of the universe. Aristotle, for one, most emphatically warned that the universe should be thought of as an orderly house but without a master, or a well-ordered army but without a commander.[23] Clearly, it could not be the idea of creation out of nothing that Aristotle found missing in the view of those according to whom 'nothing is ungenerated but everything comes to be'.[24] As to the third text, a passage from a work falsely attributed to Aristotle, the reference to the view of a coming into being out of nothing comes as a mere afterthought. One is not even certain whether the remark that the view in question was held 'not just by somebodies but by men who have a reputation of being wise' had not been made by Pseudo-Aristotle with a tongue in cheek.[25]

At any rate, these three texts cannot justify the conclusion of A. Ehrhardt that 'in the centuries after Aristotle the pagan doctrine of *creatio ex nihilo* gained some considerable ground'.[26] Such a conclusion would be questionable even if those very few texts proved that there was indeed a pagan doctrine of *creatio ex nihilo*. Any refutation of Ehrhardt's strange conclusion should start with a reference to his silence about a vast array of texts witnessing to the belief of the Greeks of old in the eternal, ungenerated, uncreated character of the world. The strangeness of Ehrhardt's conclusion comes even sharper into focus when he cites two passages from Sextus Empiricus, the famed sceptic of the sixth century of the

Christian era. But the passages, in which reference is made to coming into being out of nothing, only serve as illustration of the claim that, in line with Sextus Empiricus' radical scepticism, truth is not different from lie, and existence from non-existence.

The advent of Christianity was not enough to cure Greek thinkers of far greater merit than Sextus Empiricus of their confusion concerning the notion of coming into being out of nothing. This is illustrated by the very passage which Ehrhardt quoted in his monograph *The Beginning* as a further proof of his thesis that the *creatio ex nihilo* is not a specifically Judeo-Christian doctrine. The passage is by Atticus, a representative of the Middle Academy, who wrote that 'the cosmos was made by the noblest work by the noblest workman, who granted power to the creator [maker, *dēmiourgos*] of the universe by which [whom] he made the cosmos which previously was not'. To be sure, Ehrhardt was more aware in this case of the ambiguity of the expression and of the sceptical jeer grafted on it in some of the other cases. But he now identified the expression 'out of nothing' as the source of scepticism concerning what is real or not. He seemed to be oblivious to the fact that it was precisely the Christian doctrine of creation out of nothing which was the mainstay of Christian realism and which forced (together with the dogma of Incarnation) some patently idealist Christian philosophers to reassert the truth of flesh and blood reality in at least a roundabout way. Ehrhardt treated Christian martyrs not a whit better than he did Christian philosophers. He had, of course, to recall that the profession of a creation out of nothing by the mother of those seven brothers celebrated in the Second Book of Maccabees had become so much a part of early Christian mentality as to enter into the text of various Acts of Martyrs. In all those cases, Ehrhardt argued, one was in the presence of an outlook for which the physical reality, whether of one's body or of the cosmos, is easily to be given up because it represents a sham if not mere illusion, and precisely because it was created 'out of nothing'![27]

A Christian dogma

But were the Christians of the first three centuries, centuries of regularly recurring persecutions, unaware of the programmatic reassertions in the first chapter of Genesis that, as God said, all created things were good? Did not Christian thinkers of those centuries have their worst encounters with those, such as Gnostics and Manichaeans, for whom physical reality was either an illusion or the embodiment of evil, a sham in short? If Christians of those times seemed to deprecate the body, it was not because they rejected heaven and earth, that is, the material universe, but because they looked upon it as merely preparatory to a new heaven and earth. This basic Christian sequence was not, however, conceivable if the universe was eternal on account of the allegedly divine nature of its superlunary parts. This is why the pagan dogma of the eternity of the world prompted Christian thinkers of the first four centuries to produce a great variety of philosophical proofs of its temporal character.[28] Indeed, a reply to the question whether matter is inherently eternal or not became equivalent to a decision for or against Christian faith. Herein lies the reason for the deep-seated uneasiness voiced by such champions of classical paganism, as Jamblichus and Emperor Julian, with regard to Plato's notion that the present orderly shape of the world, though not the world itself, began in time. Far removed as such a notion could be from creation properly so-called, it was, in that famed Apostate's words, 'not without risk to assume it even as a bare hypothesis'.[29]

A further and equally revealing piece of evidence for the connection between Christianity and creationism is provided by a startling shift in the Jewish position during the first Christian centuries. On the more philosophical side the case of Philo is the best known. A contemporary of Jesus, Philo came to grips time and again with the first chapter of Genesis, but his interpretation of it showed him closer to Greek eternalism than to Biblical creationism.[30] To be sure, the latter doctrine was still explicit in the Book of Enoch, the Book of Jubilees, and in Fourth Esdras, all composed during the first century. But a few centuries later the earliest Midrashim, called 'Bereshit Rabbi', showed that Jewish theologians were no longer willing to uphold the doctrine of the complete submission of matter to the Maker of all. The situation became worse in the Jewish Cabbalistic lore where obscurantist mysticism left no opposition to cavorting with emanationism.[31]

The contrast between the uncompromising Christian resolve to voice and articulate the doctrine of creation *ex nihilo* and the growing resignation on the part of Jewish thinkers, philosophers as well as theologians, into the view in which primary matter is not created, is too strong to appear insignificant. Whatever the full explanation, it must include the Jewish resolve to oppose Christianity and whatever it specifically stood for. The centre of that resolve was Palestinian Jewry which decided to reject, toward the end of the first century, the Alexandrian list of Old Testament books, a list which had already been espoused by Christians. That list, tellingly enough, included all the deuterocanonical books, among them the Book of Wisdom and Second Maccabees, both of crucial importance for the subsequent development of the dogma of creation. No less important should seem the derisory and defamatory references to Jesus in the Talmud which obviously aim at discrediting his memory in Jewish consciousness.[32] The paramount role which belief in Jesus as the only begotten Son of God played in the Christian espousal of *creatio ex nihilo* as an article of faith can but strongly suggest that the growing indifference among Jews toward *creatio ex nihilo* may not wholly be independent of their often very negative attitude toward the Master from Nazareth.

The customary designation of the dogma of creation as a Judeo-Christian tenet raises therefore more questions than it settles. In view of the foregoing it will not perhaps appear a rank exaggeration to call the dogma of creation out of nothing, the only creation worth considering, a Christian dogma. To call that dogma Christian just because it became universally known and widely shared through the spread of Christianity, is a failure to see beneath the surface. Christianity was able to carry that dogma far and wide only through the strength provided by faith in the Incarnation. Not surprisingly, it is again the dogma of the Incarnation which helped Christians to unfold the full meaning of the dogma of creation by vindicating the true nature and dignity of created minds in the cosmos, a point, as will be seen, of crucial importance for the fate and fortunes of science. While belief in the immortality of the soul is a minor and peripheral phenomenon within Judaism, it is the very core of Christianity and precisely because of belief in the Incarnation. Without such immortality no meaning can be given to the reality of Christ between his death and resurrection, nor to the meaning of many of his words, such as for instance, 'Today *you* will be with *me* in Paradise', words spoken to the good thief facing with Jesus an imminent physical end. No wonder that from reason's probing into the mystery of Incarnation there also derived a view of man as a person with inalienable dignity.

Creator, God Incarnate, creation out of nothing, immortal soul, and human dignity are notions that form a closely knit unit, a fact well attested by the story of the dogma of creation. A milestone in that story is the Fourth Lateran Council (1214) which made the expression *ex nihilo* an official part of Christian dogma.[33] The move may appear 'a most serious mistake'[34] to anyone suspicious of philosophy in the articulation of faith. The expression *ex nihilo* is indeed very philosophical in comparison with the purely biblical phrase 'Maker of heaven and earth'. But that phrase had already proved itself ineffective to oppose efforts aimed at retaining paganism with a Christian coating. These efforts follow Christianity as a perennial shadow which bears witness to the grim resolve of human nature to oppose both Creation and Cross. In the decades preceding that Council the resolve saw a violent resurgence with Cathars, Bogomils and their kindred on the rampage. For them matter was either unreal or was evil, and in the latter case a principle on equal footing with God, the Creator. The resounding voice of a Council was clearly in order.

Another, anything but crude trend at its peak at that time represented perhaps an even greater threat to Christianity and rationality alike. Its source was the wholesale surrender of Islamic intelligentsia to the Greek world-view steeped in the dogma of the world's eternity. The chief spokesmen of that intelligentsia, Avicenna and Averroes, settled problems of the relation between revelation (faith) and philosophy (science) with recourse to the principle of triple truth. One truth, given in the plain words of the Koran, was for simple folk; another for theologians interested in distinctions; a third, or highest form of truth, for philosophers who had already found it in Aristotle. That the chief opponents of this trend were mystics – Al-Ghazzali and Al-Ashari, who rejected all philosophical approaches to points of faith, above all its chief point, Creation – shows that monotheism in its Islamic form was as incapable as it was in its Judaic version from becoming a vehicle of the dogma of creation in a way consistent with revelation, with reason, and with that notion of an orderly world which is demanded by science. It was no accident that Al-Ashari was forced to propound an atomistic notion of creation, according to which the world is created anew by Allah at every moment and that there was no causal connection between any two momentary worlds. Such worlds could in no way constitute the cosmos needed by science.[35]

In view of what happened within Islam and of the inconsistencies and hesitations which with respect to the notion of creation plague even the finest product of medieval Jewish thought, *The Guide for the Perplexed* by Moses Maimonides,[36] the story of Christian philosophy during the Middle Ages may appear to have superhuman aspects. Indeed, its best spokesmen never failed to acknowledge their indebtedness to that superhuman principle which is Christian revelation. Thomas, for one, noted time and again that Christian philosophy rests on a notion of reality which derives from the doctrine of creation out of nothing and from the latter's basis, the name Yahweh as revealed in Exodus. Other great Scholastics, such as Bonaventure and Scotus, said exactly the same.[37] Beyond the basics there were notable divergences even among these chief luminaries. Revelation did not mean a rubberstamping of rational inquiry. Nor did it prejudge the powers of reason. Yet, 'rationalist' as Scholasticism could be, and tempting as it was to conclude, against the Averroists claiming the eternity of the world, that its temporality can be demonstrated by reason, Thomas refused to be carried away. The creation of the world in time could, according to him, be known only through faith (revelation) not by philosophy, a conclusion which should give pause to some modern cosmologists and especially to many science writers caught in the fashion of estimating the past of the universe.

Creation and science

Creation in time is an evidence of God's absolute freedom to create, a freedom which Thomas always emphasized. Had he lived to see the condemnation in 1277 by Étienne Tempier, bishop of Paris, of 219 propositions, several of which could be found in his writings, he would have hardly lost his proverbial composure. He would have agreed with the instigators of the condemnation, animated by Augustinian and Franciscan voluntarism, that it was right to insist on the freedom and power of God to create any number and even an infinite variety of worlds and to make the celestial bodies move along straight lines. But Thomas would have wisely noted that too much insistence along such lines could lead to denying inherent and consistent rationality to God's creation. Such was the course of inner logic soon in full evidence. Rationality was put to flight when Ockham denied that it was not necessary to assume a necessary causal connection between stars and the light emitted by them.[38]

The denial had its roots in Ockham's nominalist distinction between God's absolute will and ordained will, a distinction which according to some recent claims played a crucial role, partly through its espousal by some early Protestant divines, in the rise of modern science.[39] The record hardly bears out this claim even when distinctly Protestant ambiences are considered, such as seventeenth century England, the Netherlands, and Northern Germany, hardly the exclusive cradles of modern science. As to earlier times, the presence of nominalism is merely nominal in the thought of Buridan and Oresme, while the presence there of a pre-Ockhamist traditional notion of creation is rather robust. There is no Ockhamism whatever in Buridan's account of the world as a huge clockwork produced out of nothing by the Creator and put in motion by him. Buridan's disciple, Oresme, and the many disciples of the latter followed suit. They made it a consensus throughout the European system of late medieval universities that, unlike the Aristotelian heavens, the heavenly clockwork, as created by God, could maintain its motions in virtue of the original 'creative' impetus imparted to it,[40] a crucial anticipation of the idea of inertial motion and one of the foundations of Newtonian physics.

The case of Copernicus, the next great figure in the rise of modern science, is equally instructive.[41] Copernicus' departure from the Aristotelian-Ptolemaic system was far greater than that of Oresme who even with respect to the earth's daily rotation remained most hesitant. But Copernicus, a Christian Platonist, was no less distant from Ockham. As a Platonist, Copernicus proceeded *a priori*. The wider scientific rationality he looked for in the universe could only come from his own mind's preferences. There was nothing in the Christian notion of God and creation to suggest heliocentrism. Its idea had to have natural roots, among them Pythagoras' dicta. But the supernatural or Christian idea of the Creator was equally indispensable that the roots might grow into a healthy plant. The growth was nourished by the confidence which derives from the Christian view of the created mind as a replica, however modest, of the divine mind. The health was provided by the awareness inherent in the no less Christian view that the Creator was not obligated to cast his creation in any particular form, be it the heliocentric system. Therefore the ultimate truth of the latter could not come from its pleasing this or that created mind, but only from its agreement with the brute facts of created nature. Finally, a recall by Copernicus of God's all-powerfulness could help make appear plausible what Copernicus' still Aristotelian physics could not do, namely, keep the heliocentric system in one piece in spite of the rapid motions of most of its members.

The role of *a priori* notions as controlled by the idea of Creation was also evident in the thought of Galileo and Newton, a point which cannot be emphasized enough. They were the ones, and not Bacon, this most ineffectual legislator of scientific method through his advocacy of a pedestrian empiricism, who created modern science. That Bacon became the patron saint of the Royal Society was a supreme irony. Newton's writings contain not a single reference to him. The irony was to continue when Hume and Kant claimed to achieve in philosophy what Newton did in physics. Of that physics both knew at best a smattering and much less were they familiar with its true epistemological foundations. No wonder that Hume, a derider of Creator and causality, ended up advocating a universe with irrational parts. As for Kant, who denied to reason access to God, soul, and the universe, he locked, with admirable consistency, everything into the subjective categories of his not too easily accessible mind.

The same century and the next, drifting steadily from traditional Christian belief in the Creator, were rich in lessons for understanding the connection of faith and scientific thought. Deism, in which the Creator was a pale image of his true self, proved no barrier to pantheism already advocated by Spinoza. It found favourable soil in German enlightenment and idealism, from Jacobi to Goethe, from Kant to Schelling, Hegel and beyond, parallel routes marked by miscomprehension of, and at times plain antagonism to science. On the French scene, Victor Hugo, who became a pantheist under the influence of Alexandre Weill, a Jewish cabbalist from Alsace, deserves special attention for having put in poignant terms that fissure and radical uncertainty which creation must evoke within pantheism: 'The always imminent end, no transition whatever between being and non-being, the return to the chasm, the slide possible at every moment, such a precipice is creation'.[42] This is a strikingly accurate description so long as the work of creation is not taken, as it has always been in the Christian context, for the work of an infinitely good Father. In Christian theology, and there alone, creation goes always hand-in-hand with conservation and this is why belief in creation functioned in history as the great leaven of confidence.

Deism was no barrier to materialism, which from d'Holbach through Diderot to Moleschott and Engels produced no serious science. It rather ended in reinstating the ancient dogma of eternal returns, that chief nemesis of the scientific enterprise in all ancient cultures, including ancient Greece. As for deism professed by Comte, the father of positivism, it was part of those preconceived limits which Comte prescribed to science, especially to astronomy and physics, limits so narrow as to constitute a straightjacket. Nietzsche, the most animated mouthpiece in the nineteenth century of the idea of eternal returns, was no less effective in discrediting himself by what he wrote on science. All these, in many ways very disparate thinkers, were at one in denouncing the Christian dogma of creation out of nothing as the worst aberration of the human mind. They certainly deserve credit for being so consistent.

The Christian response to that more than a century-long development was far from what it should have been. As for Protestant theology, it had split into two camps, one heavily pietistic, the other strongly liberal. The former professed disrespect for the rights of reason, the latter largely surrendered to its unreasonable demands. Catholic theology was in no less serious a disarray. Prior to the French Revolution it became fashionable for Catholic theologians to hail as the most effective antidote of materialism that very Descartes in whose rationalism their forebears rightly saw an invitation to materialism. (Cardinal Gerdil, an enthusiastic Cartesian, would have almost certainly been elected pope in the midst of the French Revolution, had a veto not been cast against him in the name of the Emperor of Austria at the conclave which elected Pius VI in 1800.) The upheavals of the French Revolution, rooted much more in emotions than in reason, triggered the emotional response of Romanticism. In the field of theology it appeared as traditionalism or fideism. It reflected diffidence in man's mind to the extent of tracing reason's recognition of Creator and creation to a hypothetical primitive revelation.

Those mindful of the enormous though ephemeral appeal in our times of the romantic philosophizing of Bergson or Teilhard de Chardin about creative evolution and kindred topics, will demand no special documentation of the immense popularity of the works of a Chateaubriand, de Maistre, Lamennais and others stressing the primacy of emotions over reason in matters theological. Yet popularity here too was equivalent to winning a battle by accepting grounds on which the war could only be lost.[43] It should therefore appear all the more significant that the See of Rome, Christendom's most ancient and highest authority, lost no opportunity to warn of the dangers inherent in that popularity by stressing the rights of reason. The culmination of those warnings came with the First Vatican Council which intended to review the whole range of Christian doctrine *vis-à-vis* modern trends of thought. Circumstances restricted the Council Fathers' work to two issues, one of which, the declaration of papal infallibility, invariably overshadows in popular as well as scholarly memory the other issue, not a whit less significant, the relation of reason to revelation, that is, salvation history. The Council, in line with a tradition almost two millennia old, could but insist on the very foundation of that relation which is man's ability to see the reasonability of revelation, which in turn is inconceivable if man is not able to infer from the world surrounding him the existence of its Creator. Or in the words of the Dogmatic Constitution of the First Vatican Council:

> 'The Holy Catholic Apostolic Roman Church believes and confesses that there is one true and living God, Creator and Lord of heaven and earth . . . This one only true God . . . created out of nothing, from the very first beginning of time, both the spiritual and corporeal creature . . . The same Holy Mother Church holds and teaches that God, the beginning and end of all things, may be certainly known by the natural light of human reason, by means of created things . . . but that it pleased His wisdom and bounty to reveal Himself and the eternal decrees of His will to mankind by another and a supernatural way'.[44]

As a proof, the Constitution quoted the Letter to the Hebrews (1:1–2) about God having spoken in many ways through the prophets and in the very last days through his own Son, and the Letter to the Romans (1:20) about the invisible things of God as being clearly seen from the creation of the world through the things that are made, a passage which is a direct echo of the one already quoted from the Book of Wisdom.

The wisdom of that Dogmatic Constitution is still to be appraised and assimilated by theologians, Catholic and Protestant alike. The theology of creation has been for many decades a stepchild in comparison with a large number of theological topics centring mostly on the notion of the Church. That enormously much has been written on the vast topic of the Church does not mean that all therein is now clear. Indeed, there are signs of a growing confusion which may have its source in the neglect, theological and philosophical, shown to the absolute starting-point, creation, with respect to existence as well as to any reflection on it.

Since this must be true in an unrestricted sense on account of the universal relevance of creation, there ought to be an intimate connection between the fate and fortune of science and the Christian dogma of creation. Indeed, such a connection is all too well attested by a history of science free of the shackles of the blind rationality of logical positivism and of the subtle irrationalism of a psychologism and sociologism erected into epistemology and metaphysics. With that history of science most theologians and educated Christian laymen still have to achieve a modest familiarity. Less inexcusable is the now fashionable slighting of a development in philosophy which in a sense was a response to that Dogmatic Constitution. The call by Leo XIII for a return to the Christian philosophy of Thomas Aquinas, a philosophy of realism, was inevitable because the recognition by the human mind of the existence of the Creator by means of created things implies a philosophical framework which makes no sense unless there is a totality of things, real, orderly, and contingent, which is the notion of a universe proclaimed by the dogma of creation and claimed by science as well.

CHAPTER FOUR

A BOOKISH PHILOSOPHY

Books as things

Among philosophical systems realism is only one among many whose number keeps steadily increasing. No less steadfastly do their respective spokesmen charge one another of being in basic error. This disagreement is even true of systems which carry the label realist. The realism which claims to reach from the consideration of real physical things the highest level of metaphysical reality or God, has little in common with the realism of a so-called 'rational metaphysics', which wants to forestall that very reach. Competing and mutually exclusive philosophical systems readily give the impression of a cacophony which can but invite condescending remarks on the part of those whose expertise was, not too long ago, called natural philosophy. Science was still an upstart with respect to philosophy when Galileo declared that whereas the conclusions of science 'are true and necessary and have nothing to do with human will', in matters of law and humanities (philosophy above all) 'neither true nor false exists'.[1]

Such a remark seems to imply that science needs no philosophy, or if it does, it provides its own, a philosophy immune from the confusion characteristic of philosophies. However that may be, the disarray in which philosophical systems find themselves may suggest that the only thing they have in common is that they are published in books. To be sure, there were philosophers and philosophies before books and even before writing. As the symbolizing of thought in phonemes, or spoken words, had long preceded its registering in visual symbols, philosophical thought was first carried by oral tradition, which, by being so palpable a matter, is hardly less concrete than a script. The advent of printing was merely a powerful manner of spreading symbolized thought far and wide. After five hundred years of printing the manner is ordinary, nay trivial. The statement that philosophies have in common only their being set forth in books will certainly appear trivial to those who, unlike Gutenberg's contemporaries, experience not even a touch of wonder while holding a book in their hands. That wonder was so great at that time as to make some think that Gutenberg was consorting with the devil. There may seem indeed something devilish in man's ability to hold his own thoughts in his hands and to make those thoughts practically independent of the constraints of space and time. Books seem to take on an existence of their own, which prompted Popper and some others to see in books a third world, on equal footing with the world of mind and the world of matter.[2]

Such an importance given to books and other symbolic recordings of thought will appear an extreme to anyone mindful of the difference between maker and product. At any rate, if the mind of an author is on the same footing with its book-form expression, he is clearly barred from defending his own mind and meaning from the sundry misrepresentations to which any book of some scope invariably gives rise. The other extreme, the slighting of the philosophical significance of books, of any book however trivial, is no less self-defeating. Since any book is a tangible product and obviously made for a purpose, any book written either against tangible reality or against the reality of purpose is the very refutation of its author's claim. But a book is also a refutation of the slighting of the excellence of thinking, that age-old citadel of metaphysics. Materialists and positivists, be they logical or not, who must exorcize even the most rudimentary form of metaphysics which is embodied in thinking about any physical thing, are also refuted by the very books they write. A book is a thing, if it is anything. But a thing, any thing, is so loaded with metaphysical realism that, tellingly enough, Wittgenstein's second statement in his *Tractatus Logico-Philosophicus* is the defensive declaration that 'the world is the totality of facts not of things'.[3]

Yet the *Tractatus* was as much a thing as it was a fact, though hardly the end of philosophy in the sense in which Bertrand Russell greeted its publication in 1921.[4] Far from reaching the very farthest point on the road possible for a certain philosophy, a road ending where metaphysics begins, the very title of Wittgenstein's book spelled the death knell on that kind of philosophy. The word 'logical' qualifying philosophy once more proved to be a Trojan horse. Philosophy, as any other reasonable human activity, must of course be a logical, that is, consistent enterprise. But if philosophy is logical by emphasis, it is no longer that love of wisdom which philosophy is by hallowed appellation. Whatever the wisdom of logic (professional logicians, to say nothing of students of logic, rarely give the impression of being men of wisdom), logic operates within its self-imposed conceptual circle and is very different from that love the aim of which is to reach a reality distinct from the self. A philosopher, who programmatically encloses himself within the circle of the logical analysis of a set of concepts, will know neither of love nor of wisdom, nor can he claim, if he remains logical, that is, consistent, the attention of other philosophers. The logician, or the philosopher who is logical by emphasis, is an advocate of solipsism. All he can do with it is 'to cherish it',[5] the devastating advice given by Chesterton to a Cambridge undergraduate, who boasted of his conviction about his own existence as the only reality which he could hold with certainty.

Most philosophers who fall in the category of solipsists do so by the implications of their postulates rather than by conscious choice. They not only can have no true love, but they cannot even wonder, an act in which Aristotle rightly saw, at the very start of his *Metaphysics*, the beginning of philosophy. As is well known, Wittgenstein was on occasion seized with the strange feeling of wonder why anything exists at all,[6] but this was the very sentiment which the *Tractatus* meant to banish from philosophy by making it 'logical'. Much the same, by the way, is the aim of the advocates of a philosophy called 'scientific'. Philosophy, according to the *Tractatus*, is the account of a state of affairs which should appear rather uninviting of any wonderment: a proposition and that which it describes must have the same logical form or structure. Revealingly, Wittgenstein did not feel a sense of wonder but a rude shock when a friend, Piero Sraffa, an Italian economist working in Cambridge, made a gesture of contempt and asked: What is the logical form of that?[7] Unfortunately, the gesture forced Wittgenstein to go from bad to worse, from the tautology of logicism to game theory. In the former, reality could not be reached though one could still seriously think of it; in the latter, it was a mere plaything, a very casual matter. Something of this was foreshadowed by the very first phrase of the *Tractatus*: 'The World is everything that is the case'.[8] Why his own book was still meant to be not merely a case, or casual fact, but a tangible and purposeful embodiment of a specific and universal proposition about any conceivable case, Wittgenstein failed to consider.

A strange failure indeed, but in a sense understandable. Wittgenstein, once a student of Russell, could not have learned from that major monograph on Leibniz, which signalled Russell's entry into the field of philosophy,[9] about Leibniz's proof of the existence of God based on the existence of books.

In an essay written in 1697 on the ultimate origination of things, an essay which remained unpublished until 1840, Leibniz started, of course, in an emphatically un-Wittgensteinian way by declaring that the world was an 'aggregate of finite *things*' (Italics added). Whatever Wittgenstein's disagreement on this point may have been, if his book truly was the truth about any and all cases the world may be, he could not have 'logically' taken exception to Leibniz's next statement that the aggregate in question was ruled by a unity which gave to that aggregate its specificity and coherence, and 'created' therefore in a sense the world. To illustrate the point Leibniz referred to books:

> 'Suppose a book on the elements of geometry to have been eternal and that others had been successively copied from it, it is evident that, although we might account for the present book by the book which was its model, we could nevertheless never, by assuming any number of books whatever, reach a perfect reason for them; for we may always wonder why such books have existed from all time; that is, why books exist at all and why they are thus written. What is true of books is also true of the different states of the world, for in spite of certain laws of change a succeeding state is in a certain way only a copy of the preceding, and to whatever anterior state you may go back you will never find there a complete reason why there is any world at all, and why this world rather than some other. And even if you imagine the world eternal, nevertheless since you posit nothing but a succession of states, and as you find a sufficient reason for them in none of them whatsoever, and as any number of them whatever does not aid you in giving a reason for them, it is evident that the reason must be sought elsewhere . . . [and that

> therefore] . . . even by supposing the eternity of the world, an ultimate extramundane reason of things, or God, cannot be escaped'.[10]

As an author of books, Wittgenstein should have logically asked the questions posed by Leibniz, but as the author of a 'logico-philosophical' treatise, he barred himself from raising them. None of the questions raised in that Leibnizian text and context – questions about the specificity of things, about their making a coherent unity or universe, about the merit of infinite regress – make, of course, sense if no assent had already been given to the existence of particular things, be they mere books, an assent demanding more than 'pure' logic. Leibniz himself was hampered in that respect. Insofar as he was a rationalist he could not prove that man knew reality.

A novel starting point

Leibniz, as is well known, resolved to his own satisfaction the existence of things and their coherent interaction by a recourse to the principle of pre-established harmony, that is, a God-ordained co-ordination between things and thought. Such a pre-established harmony satisfied practically nobody but Leibniz and the Leibnizians. Others knew all too well that Leibniz's was the second heroic though futile attempt to recover external reality which Descartes had already pre-empted of its meaning by making the *cogito* the starting point of philosophy. Descartes' immediate disciples had indeed realized that the principle of 'clear and distinct' ideas, on which Descartes based everything else, deprived him of the possibility that the mind may be the beneficiary of the bodily sensations, because interaction was inconceivable between two so radically different entities as mind and body were according to Descartes. Malebranche, a most acute and enthusiastic

Cartesian, tried to shore up matters by postulating occasionalism, a doctrine according to which all interactions between all physical things, not only between mind and body, are inscrutable. Any change in our body as evidenced by sensations is, Malebranche argued, an occasion for God to produce a corresponding change in our mind.

Such was a mere atomization of the pre-established harmony advocated by Leibniz. Both procedures put man's freedom in jeopardy in addition to depriving him of his ability to go from himself to the external reality. Typically, the monads of Leibniz were windowless entities and they were of no help in opposing the criticism which occasionalism received at the hands of Berkeley, who concluded that it was better to abandon all concern for external reality and to state that only the mind existed. The mind in turn saw, so Berkeley believed, all things in God. Such was one possible terminal status of Descartes' initiative. Another was Spinoza's pantheism in which difference between mind and body was completely dissolved into the universality of nature which for Spinoza meant God himself. Such a God had to produce the world since eternity. If God was infinitely perfect, Spinoza argued, an infinite number of things in an infinite number of ways followed from the necessity of divine nature with that necessity with which in a triangle the 'three interior angles are equal to two right angles'.[11] The argument is from the *Ethica ordine geometrico demonstrata,* possibly the most mistaken enterprise ever undertaken by a philosopher. Being tautologies, the demonstrations of geometry cannot derive, in obvious rebuttal of the claim that an infinite number of things exist in an infinite number of ways, even the specificity of a single thing, let alone of their totality, the universe. Indeed, Spinoza himself recognized, shortly before his death, that the existence of concrete finite things is well nigh impossible in the system advocated in the *Ethica.*[12] What the glorification of the infinite in the *Ethica* had anticipated was the infinite distance to which any 'logical' or 'scientific' philosophy removes itself, in proof of its inanity, from concrete physical reality on which it purportedly stakes its fortune. The same philosophy, as was the case with the *Ethica,* is no less removed from the less tangible though no less real realm of ethical behaviour and norms.

Much less known than the dicta of the *Ethica* is a statement of Spinoza, a succinct survey of the chief choices available in philosophy: 'The Scholastics start from things, Descartes from thought, I start from God'.[13] Clearly, the Scholastics were not entirely forgotten in 1671, only the prevailing fashion prevented the age of reason from seeing anything reasonable, let alone the only reasonable factor that secures reality for man, the recognition of the existence of things, the starting point of the great Scholastics. As will be seen later, that starting point had only a superficial resemblance to the one which was tried by Locke who declared the mind a *tabula rasa* and postulated the radical priority of sensations. Anyone aware of the difference between sensations felt by the subject and things, or objects, that give rise to sensations, will easily realize that Locke's starting point was merely a dialectical rewording of the Cartesian stance. The reasoning of Locke could provide no more assurance about the existence of things supposedly activating the senses acting on the mind than could the system of Descartes make appear plausible the mind's grasp of sensations, let alone of things.[14] Hume unfolded the logic of all this by declaring the lack of objective validity in the so-called principle of causation. Since the latter was reduced by Hume to the status of mere habit, no better was the status of all dicta concerning either the chain of causation along which the mind reached the First Cause, or the creative causation through which God calls forth the world. All those dicta, so Hume declared, expressed habits which, as is the case with all habits, are mere conveniences and fashions, and can never strictly assure us of the existence of objective reality, let alone of its metaphysical kind.[15]

Such was the voice of the complete sceptic which Kant tried to stifle by arguing over the 600 pages of a laborious book that causality is a necessary imposition of a category of reason on the phenomena by which he meant sensory impressions. The absolute priority of mind over things, this starting point of Descartes, received thereby its most radical form, but no less complete was now the impossibility of knowing things or noumena. Kant did not suspect that any moderately critical reader of his *Kritik* would wonder why its author spoke profusely of noumena which were in principle unknowable. Kant himself thought that he had forestalled such an eventuality by postulating the practical reason through which not only things (noumena) but also their totality, the universe, could be reached, in addition to the soul and God as well. It shows something of the ineptness of the practical reason that the finest adepts of Kantian philosophy, Fichte, Schelling, and Hegel, looked for a road to real things through speculative reason, that is, metaphysics, which they thought to exist beneath Kant's emasculation of it.[16] The road specified by them obeyed, as had to be the case with great thinkers, the inner logic of initial presuppositions. They brought out only more strongly the utter subjectivity of a mind isolated by Kant from external reality.

A disastrous end

The subjectivism of German idealism accounts for its hostility to science, a hostility subtly present even in French positivism.

As first articulated by Comte, it was called to replace concern, a clearly metaphysical one, about reality as such. Science within such a perspective cannot be more than an economic correlation of sensory data, a procedure intrinsically free of any need of a correlating mind. Mach himself declared that if all sensations were available to man, they would naturally correlate themselves in the most economic way and render science thereby unnecessary.[17] Such was a radical dismissal of reality and mind in the name of science which received its sharpest rebuff at the hands of science. As Planck and Einstein, both erstwhile admirers and professed disciples of Mach, gradually realized, their creativity in science implied a philosophical outlook very different from any variation of idealism, even if it be cloaked in the praise of sensations.

Planck and Einstein remained lonely voices. The dominating trend of the philosophical interpretation of modern science was set by Bohr, who took Höffding's pragmatist complementarity as a solution to some philosophical problems posed in atomic physics. The hallmark of that pragmatism was the resolute avoidance of any question about reality as such. That in atomic physics the wave and particle aspects were equally useful, though in a complementary way, was certainly true. But was this a justification for Bohr's warnings against questions concerning reality and what is always implied in it, causality? Again, it was true, as first recognized by Heisenberg, that the combination of quanta and wave functions made impossible in principle the simultaneous measurement with perfect accuracy of pairs of canonical conjugates, that is, to mention only two such pairs, momentum and position, energy and time. But was it philosophical to argue that what could not be measured exactly, could not exist and take place exactly? Was it reasonable to argue, to take only one concrete example, that since the exact time of the radioactive disintegration of a single atom could not be predicted and measured exactly with the tools of theoretical and experimental physics, the transition of that atom from one stage to another evidenced by its radioactivity was not caused by anything in any sense?

It shows something of the philosophical poverty of a richly scientific age that such a disregard of a question bearing on the very core of reality was taken to be very philosophical, that is, reasonable and rigorous, only because it rhymed with some technicalities in science. One may, however, suspect that belief in the 'scientific' overthrow of the principle of causality would not have become a tone of thought of our age, had this scientific age of ours not already parted with belief in the First Cause which makes all other causation possible. The 'scientific' discredit brought on reality and causality did not spare science itself. It soon became fashionable to philosophize about science in terms of sociology, psychology, and all sorts of behavioural subsciences in complete disregard of ontology and metaphysics. The result was a parcelling of science into disconnected revolutions, paradigms, images, research programmes and the like. Those who know something of the remorseless law of logic will not be surprised on finding that if reality or objective coherence fails to be accepted, any analysis of knowledge becomes a celebration of incoherence. The latest and perhaps ultimate outcome along the line initiated by Bohr and the Copenhagen School concerning the interpretation of our knowledge of physical reality is the so-called multiworld theory according to which there are as many worlds as there are observers.[18] The coherence of worlds and observers is, within that theory, a matter of wishful thinking at best. The Cartesian-Kantian precept of the primacy of the mind and/or observer may not appear overly harmful when this or that prominent scientist claims that his observations constitute reality. The claim is largely pre-empted of its meaning by the complicated instruments at his use, instruments not fashioned to his whims and fancy. But when everybody, with instruments or not, can claim to constitute with his observations the world, will there still be a cosmos against which theories can be tested in a universally valid and objective manner? Einstein's warning about the Copenhagen people – 'what a dangerous game they play with reality!'[19] – reveals in retrospect a prophetic quality.

That the multiworld-theory provoked no loud dissent and outcry among cultivators of quantum mechanics, that many of them greeted the evasiveness of Taoist dicta as an answer to basic philosophical problems of modern physics – these and similar symptoms evidence the extent and depth to which diffidence about reality dominates the contemporary reflections on science. This is a point of utmost importance for anyone pondering the correlation of cosmos and Creator. For this correlation to be effective, there ought to be an effective acceptance of a coherent reality existing independently of one's thinking. A philosophical theist can therefore derive comfort from the fact that the same diffidence about reality which blocks the rational way to God at its very start, puts science too on a road which is a mere blind alley, and into which not a few scientists glibly march, though not by their deeds but by their inept philosophizing about them.

True common sense

The extent of that diffidence is so great as to assert itself even among those whose forebears, to recall a remark of Spinoza about the Scholastics, started from things. It is not at all generally believed in Scholastic circles today that one must start with real things in order to assure a consistent knowledge. Sometime before it became a futile fashion among some self-styled Thomists to transcend Thomas by transcending

Kant, as if lifting oneself by one's bootstraps were possible, Neo-Scholastics were plagued by a recourse to the term 'common sense' as if there was a special faculty of the mind through which immediate access is given to reality.[20] The recourse was in part necessitated as a tactic to neutralize Lamennais and others who in their Romanticism tried to ground Christian philosophy in a perception of truth and reality akin to a mystical experience. They called it the operation of the common sense, a term already given wide currency by such philosophers of the Scottish school as Reid, Beattie, and Stewart. Currency did not mean necessarily that solvency which is clarity for philosophy. Stewart was forced to admit that long philosophical use did not make the term 'common sense' any clearer than 'mother wit',[21] a very important gift but not too philosophical. The term, although very familiar to Latin medievals from Cicero, did not find its way into Thomas' writings. Neo-Thomists fond of the term, if only to pre-empt the position taken by Lamennais, had no small trouble in grafting it on Thomas. Pierre Duhem, a methodical positivist but an instinctive realist, who viewed himself as 'an apostle of common sense',[22] would have found it very difficult to find a satisfactory, let alone a commonly accepted definition of it in manuals of Scholastic philosophy. It was one thing to say, as Duhem did, that mathematical physics consisted in casting quantitative data into a system which only common sense could relate to reality. It was another thing to specify, which Duhem never tried, that common sense as some part of man's cognitive facutly which would provide such commonsense judgments as 'I am the same person today as tomorrow', 'The material world has an existence independent of the observer', 'The future course of physical events will resemble the past', 'There are other intelligent beings beside myself', and the like.

Whatever the possibility of such a specification, it is not found in Thomas who did not parcel the mind into various faculties. Following Aristotle, he spoke of 'common doctrines' or principles which underlie all judgments of the mind.[23] Among such principles are the tenets that everything must be either affirmed or denied, and that no thing can exist and not exist at the same time. Such principles are an articulation of the principle of identity and contradiction, and not a grasp of reality itself, be it a commonsense grasp. That principle is inherent in all the foregoing examples of commonsense truths about reality, all of which represent something distinctly specific and therefore not the most fundamental aspect of man's intelligent grasp of reality. That grasp relates to the very existence of material beings revealing themselves as sensory objects. This is why Thomas insisted: 'That which is first known by the human mind is such an object . . .; the act by which the object is known is known secondarily'.[24]

Long before Descartes posited the false priority of knowledge as such over knowing outside reality, and long before some scientists, under the impact of that priority, decided to constitute the world by their knowing, Thomas voiced a very different theory of knowledge. It was in fact not at all a theory, that is, an epistemology (a discipline always tainted with the primacy of the Cartesian-Kantian *cogito*), but a metaphysics. It is a metaphysics because, contrary to the fundamental tenets of empiricism, in the knowledge of reality not what is merely sensory is revealed but the intelligible nature of physical entities. Such a knowledge is the grasp of the being of things which underlies the grasp of any specificity evidenced by them. While such specificities are always subject to reappraisal and refinement, that is, criticism, the latter can only be carried out by a reassertion of the truth of real beings and therefore criticism cannot be the first word in philosophy. Critical philosophy works only in spite of itself.

Part of the truth of Thomas' metaphysical realism, the only proper label of a genuinely Thomistic 'epistemology', lies in its consistency. This will not appear a small matter if one recalls Chesterton's poignant observation that 'No sceptics work sceptically; no fatalists work fatalistically; . . . no materialist, who thinks his mind was made up for him by mud and blood and heredity, has any hesitation in making up his mind'.[25] Had Chesterton been a professional philosopher and had he lived today when philosophy is in many quarters a respectable enterprise only when it deals with scientific knowledge, he would have easily found some highly acclaimed targets to make his point. Clearly, the falsification theory of knowledge was not proposed to declare that theory to be intrinsically falsifiable. The process theory of knowledge and existence clearly claims a permanence while it subjects everything else to endless transformations. The theory of Gestalt switches obviously wants to retain a permanent image of itself, while it turns all other viewpoints into the prey of unpredictable sudden changes. The theory of knowledge based on the succession of basically disconnected scientific paradigms, brought about by scientific revolutions, carries its own refutation by claiming that there is a connection or structure underlying all revolutions.

The latter theory has at least the merit of having been carried by its author to its logical end where it is no longer necessary to assume that the world science deals with is an ordered entity, a consistent construct.[26] That such a world cannot logically prompt that well-ordered knowledge which is science shows the intimate connection between cosmology (science) and epistemology (metaphysics), and also something of the soundness of Thomas' starting principle that it is the existing beings which elicit knowledge. The principle is not only proven sound by that laboratory which is the history of philosophy, but also helps explain why the brute facts of nature can shock the scientific mind to such an extent as to spark profound insights about the workings and structure of the physical world to be tested in laboratories where one looks for real things and not merely for one's thought about them. Moreover, when the existence of beings grounds knowledge, there is a natural and legitimate way to pose not only the scientific questions as to the specific manner in which things exist, but also the more fundamental, non-scientific, questions: What is it to exist? In what does existence consist? Why is it that nothing that makes itself known seems to contain the sufficient reason of its existence? Why is it that whatever is known happens to be, that is, evidences a change and thereby demonstrates its contingency? Must not contingent existences have their ground in a necessary being?

Science, reality and God

None of these questions can be raised by the method of science, nor can its method provide an answer to them. The formulae of mathematical physics are equations, that is, identity relations,[27] and as such they represent a framework which cannot accommodate that very happening which at every moment generates existential novelty in the world. Those equations are no proof whatever of novelty, a point of utmost importance in view of the steady expropriation for the past 300 years of the notion of proof by mathematical formulae. Those formulae, nay the entire mathematics, are, as Bertrand Russell reluctantly came to admit, mere tautologies.[28] Pythagoras' theorem is called the bridge of dunces (*pons asinorum*) for a reason far better than what can be intimated in schoolmasters' asides. Being a tautology, as all theorems of Euclid are, that theorem can prove itself to be a bridge only for those who, being dunces, fancy that it leads to another shore, that is, to something tangibly different. Only a genuine proof can do this and therefore it must be more than a mathematical tautology, exceedingly instructive as such tautologies can be.

The proof of the existence of God based on the consideration of the cosmos is made up of steps each of which leads to something different. No such a procedure will be palatable to anyone whose mental vision has already become conditioned to the 'proofs' used in mathematics and logic, and therefore is insensitive to proofs of reality itself. The procedure will present no difficulty to anyone who recognizes that it is the beings that make knowledge possible, that every step of real knowledge is a step to something different, and that the proof of such steps is that knowledge which is to know things and to know them to be known. Once this elemental measure of realism has been mustered, no perplexity will be caused by the question of infinite regress. Within the perspective of that realism not only the difference between numbers and things will be readily seen, but also the fact that the adding of links, even if it should be in infinite numbers, to the chain, bears in no way on the chain's existence.

The world is a chain of things about which modern science has unfolded an astonishing measure of specificity, consistency, and unity. Of these three the specificity has been most within the reach of the layman throughout the whole development of science. The visible universe was never akin to that homogeneous monotony to which some *a priori* thinkers tried to reduce it in order to make it appear a necessary form of existence. Both the alleged homogeneity of the Aristotelian spherical heavens and of the three-dimensional infinity, which succeeded it, could be called into serious doubt by a mere look at the Milky Way. Its irregular contours evidenced something very specific and singular about the cosmos. No wonder that Aristotle chose to talk it away from the sky, whereas the Cartesians and Newtonians preferred to ignore it.[29] One need not, however, watch the sky in order to convince oneself about the specificity of existence. Observation of one's immediate surroundings should be enough to reveal that specificity bordering on queerness. But the specificity, the queerness of things and of their interconnectedness, the peculiar shape of the links in the chain and of their arrangement in it will be an effective pointer to its contingency only for those who have never engaged themselves in casting sophisticated doubts on reality. Only they will find remarkable the words put by the poet in Newton's mouth:

> 'Tis not the lack of links within the chain
> From cause to cause, but that the chain exists,
> That's the unfathomable mystery,
> The one unquestioned miracle that we know
> Implying every attribute of God'.[30]

To exist or not to exist is a far deeper alternative than to be alive or dead, the relatively superficial sense given by Hamlet to the phrase, to be or not to be, the very alternative which unfolds the depths of metaphysics. The gaping of those depths is inexorable. All those who try to turn metaphysics into a relic of the past by the strategy, according to which 'animal instinct is fixed reason, and human reason is mobile instinct',[31] fail to show that it is instinct and not reason which enables man to attach any meaning to that small but weighty word, *is*. It is an evidence of man's animality that many modern philosophers fail to be provoked by that small word. It takes the philosophical sensitivity of a great artist to write as Sartre did: 'Existence is not something which lets itself be thought of from a distance; it must invade you suddenly, master you, weigh heavily on your heart like a great motionless beast – or else there is nothing more at all'.[32] These words would do much credit to all those philosophers who, ever since the *cogito* replaced the *esse* as the starting point of philosophy, try to keep the question of existence at a safe distance. Instead of attaching a central meaning to that little but most weighty verb *is*, they avoid other verbs as well, in order to be free for a clever game with nouns. Nouns dissociated from verbs are not images of things, that is, of reality. They are mere concepts that act in no substantive way whatever either among themselves or on the perceiving mind. Concepts are a perfect means to reserve the field of rational discourse to mere accidents.

Such is the genesis of that global accident that has been playing havoc for some time with modern philosophy. In that accident the world itself turns into a chaotic pile of disconnected events, 'all spots and jumps, without unity, without continuity, without coherence or orderliness or any of the other properties that governesses love'.[33] In stating this in 1931 about the universe as his most fundamental belief, Bertrand Russell could not claim even that modicum of excuse which he could in 1914 when Einstein had not yet given to General Relativity its cosmological capstone, which is the scientifically consistent unity of all interacting things. Then Russell blandly asserted that such a unity is a 'mere relic of pre-Copernican astronomy'.[34] Russell, who time and again relished the role of being a highly applauded spokesman for the wider public about the latest in atomic physics and relativity, must obviously have been led by a non-scientific trend of thought in advocating the idea of an incoherent universe. Once caught in that trend anything can happen, the big turn into small, the heavy into light, the bright into dark, and even that heavily oppressive experience of existence, as described by Sartre, into a feeling of floating freely in the air, as if one were no longer a contingent being. Not that Sartre failed to warn against the delusion of denying the reality of contingence. But in the same breath he turned it into 'the absolute perfect free gift'.[35] He did so by turning logic inside out through the claim that contingency cannot be overcome 'by inventing a necessary causal being . . . for no necessary being can explain existence'.

Most philosophers uneasy about contingency would not go into such mental acrobatics indicative of a state of excessive exhilaration. They rather warn against taking the feeling of contingency too seriously, lest by looking at experienced existence as a mystery, that is, something which does not explain itself, one should look 'for an explanation of the universe outside the universe itself'. To this they add the fatherly reminder that 'Immanuel Kant gave reasons why it must be a mistake to look for something beyond, which would explain the fact of existence'.[36] Myths have a long way of dying. So many years after the fallacy of Kant's criticism of the cosmological argument had been exposed,[37] one can still sound learned, in some circles at least, by being unaware of this result.

Awareness of that fallacy does not, of course, block one avenue of escape, the recourse to relativism and agnosticism, against which there is no arguing. What can one say to a philosopher who on the one hand confesses that the existence of anything seems to him to be 'a matter for the deepest awe', but on the other hand declares that very awe to be a matter of individual preference?[38] Does not such relativism spell the death-knell of philosophy or of any rational investigation by pre-empting the recognition of problems that should be considered to be of utmost depth by all concerned? Is not the end of rationality in sight when agnosticism is the reply concerning the nature of the question: 'Why should anything exist at all?' Those aware of the fact that relativism and agnosticism are not signs of intellectual tolerance but wilful postures, will not be surprised on finding that the same philosopher decried in the same breath 'the absurd rçquest for the nonsensical postulation of a logically necessary being' as an answer to that question. It is obviously the sign of wilfulness if not incoherence when logic is suddenly jettisoned from the ship of a purportedly 'scientific' philosophy lest the ship reach homeport, that is, the Ultımate.

To steer clear of such incoherence one has to adopt a definition of rationality far wider and deeper than what is provided by a logicism so shallow and narrow as to prevent its devotees from eating in order to remain fully logical. Long before Bertrand Russell noted that rationality restricted to logic fails to provide ground for the truth of the most elementary induction dealing with such real matters as whether one's next meal would be nourishing or whether the sun would rise tomorrow,[39] Spinoza served evidence that a philosophy admitting only the absolute truths of logic cannot deal with reality: 'A man would perish of hunger and thirst, if he refused to eat and drink, till he had obtained positive proof that food and drink would be good for him'. Russell was again anticipated as Spinoza went on: 'But in philosophical reflection this is not so. On the contrary, we must take care not to admit as true anything, which is only probable. For when one falsity has been let in, infinite others follow'.[40]

Actually the very opposite is true. Once one truth, the truth of knowing reality is not let in (be it on account of its allegedly being but probable), no truth whatever about reality is any longer forthcoming and certainly not the most important truth about the totality of real things, namely, that a specific, consistent, thoroughly one universe cried out for a necessary Being as its sole *raison d'être.* Only such a Being can call into existence the universe, a process called creation out of nothing. Such a process is the deepest, though most luminous mystery which alone can shed light on anything else and prevent rational discourse from relapsing into irrational reversals of its progress. That modern science displays with astonishing effectiveness the specificity, consistency, and unity of the universe, and that on account of Gödel's theorem no *a priori* derivation of the basic structure of the universe is possible, can certainly reassure the theist that in addition to philosophy, science too is on his side and not on the side of atheists. The latter will seek refuge, as they invariably did in the past, in the alleged eternity of the universe, a point which philosophy and science can neither prove nor disprove. Assertion of the eternity of the universe is, as was aptly noted more than two centuries ago, the first article 'of secular faith'.[41] It is that faith, not science, which is heard when a prominent cosmologist declares that modern science has proved the eternity of matter and that this is its most outstanding achievement. Perhaps not even that faith is at play when the cosmologist has to address an international conference under the watchful eyes of the emissaries of the Party and if he is not made of the stuff of which heroes and martyrs are made.[42]

Natural theology

The recognition, however unequivocal, of the existence of a Creator from the contemplation of nature, that is, from the book of nature, gives one an image of God which is very bookish and pale in comparison with his image set forth in the Books of Revelation. Undoubtedly, that pale image does not have the attractiveness which produces martyrs and heroes. But the cosmological argument and natural theology, of which it forms a central part, was needed by them in the measure in which revelation had to be defended against the charge that it rested on 'blind faith'. In the history of Christendom that need was felt most acutely after the onset of Cartesian rationalism which refused to accord rationality to any argument that did not resemble a proposition of geometry. Natural theology as cultivated by Mersenne, Gassendi, Boyle, Grew, Derham, Clarke, Niewentijdt and many others reflected a consensus shared by Christian men of science of varied denominations in the soundness of the enterprise.

What they have also shared was a rather superficial knowledge of what Christian philosophy, the basis of natural theology, was about. Clarke, for instance, decried the Scholastic definition of God as *actus purus* or pure act of existence, as empty verbiage characteristic of bookish learning. He did so, tellingly enough, in a book on natural theology in which the word contingent occurred only once.[43] This is not to suggest that had Clarke read the *Summa* of Thomas, he would have done differently. Great classics, to say nothing of lesser books, serve for most of their readers not so much as a deep source of thought, but as so many mirrors of their own shallow thoughts. Apart from his lack of sympathy for 'Romish centuries', Clarke would have read the philosophy scattered throughout the *Summa* with Cartesian eyes and would have hardly noticed its gems. In all appearance they were not noticed by Mersenne, a saintly friar, who had used the *Summa* for a textbook in the various courses he had taught for decades. One wonders what Thomas would have felt on seeing his argument (the Third Way) being thrown by Mersenne into a potpourri of thirty-six ways of proving the existence of God.[44] Their variety was not a sign of richness but of a repetitous confusion which in a sense was promoted by Thomas himself.

Always intent on putting Aristotle in the most favourable light, Thomas was hardly the one to notice a glaring inconsistency in Aristotle's thought. Contrary to his stated principles, according to which a method of inquiry had a broader validity the broader its respective subject matter was,

Aristotle took the method of the investigation of the living as the model of what was valid in the study of the non-living, although the latter realm was much broader than the former. It was this biologization of physics, based on the allegedly universal applicability of the notion of purpose wherever motion was in evidence, which for almost two millennia prevented the rise of the study of the physical realm as a purely mechanical system in which questions about purpose are irrelevant. While the rise of mechanistic physics and the rise of philosophies (Cartesianism and empiricism) accompanying it, cast doubt on any and all purpose, purpose was not something to be easily cast aside. In fact, precisely because those philosophies were unable to do justice to purposeful action, purpose which is inseparable from theism, reasserted itself with a vengeance, and did so, not unexpectedly, through natural theology. An unjudicious celebration of the countless mechanical patterns, revealed by the new physics, as so many arrangements made for the benefit of man was indeed the long and short of natural theology as cultivated in the first half of the eighteenth century. Such an enterprise, short on genuine philosophy but long on pleasing illusions, could only bring discredit upon itself. Its demise was no less hastened by its severance, largely under the impact of deism, from revealed theology. The idea of a First cause, reduced to that of a disinterested cosmic clockmaker, was no barrier to materialism. The process was completed within one long lifetime. When there appeared in 1745 the first volume of Buffon's famed *Histoire naturelle*, a beautiful etching[45] showed the Creator launching the comet whose close collision with the sun was supposed to trigger, according to Buffon, the formation of the planetary system, a preliminary to the formation of the earth and the appearance on it of plants and animals. The ageing Buffon was, however, reported as having had come to the belief that attraction and repulsion could very well replace every reference to the Creator in his *magnum opus.*[46]

He would be greatly shocked today on finding that attraction and repulsion have become the hallmark of a very bookish physics, outmoded for some decades. When still before the First World War modern physics was not yet in the books, Chesterton could aptly note that modern science 'was moving towards the supernatural with the rapidity of a railway train'.[47] Today, in view of the expanding universe, a reference to the speed of light would be more appropriate. To see the situation in such a perspective demands not so much science, be it the most modern, as a philosophy with endurance, the touchstone of *philosophia perennis.* With such a philosophy on hand one can also discern the true merit of the most fashionable trend of thought in this space age of ours. The trend consists in the physicalisation of all biology, psychology, and philosophy. It culminates in heedless assertions of the presence of life and intelligence everywhere in the universe. The aim of this trend is the severing of the cosmos from its Creator by presenting the universe as a home which both makes man and is made by man. Such a game with logic is, as one may suspect, a most effective way of turning the cosmos into a trap, hardly the most beautiful of constructs.

CHAPTER FIVE

A TRAP OR A HOME?

Ethics and evolution

In the proliferation of books written during much of the seventeenth and in the first half of the eighteenth century on natural theology, the most memorable is the theme that science provided vast evidence on behalf of the Psalmist who sang of the excellence of the dwelling place, this earth, prepared by the Creator for man in the very centre of the universe. No less well remembered is the discredit brought upon that natural theology on account of its claims about anthropocentric purposiveness evident everywhere in the universe. Darwinism and astronomy were all too effective in shattering the belief that man's surroundings were made especially for him, that he was the centre of the universe, let alone that he was in its very centre, as if it were his home. It should therefore come as a surprise that a major monograph and textbook on gravitation and cosmology was, only a few years ago, dedicated 'to our fellow citizens who for love of truth, take from their own wants, by taxes and gifts, and now and then send forth one of themselves as dedicated servant, to forward the search in the mysteries and marvelous simplicities of this strange and beautiful universe, *our home*' (emphasis added).[1]

To call a place, a structure, an environment one's home is a valuational judgment, the very kind of judgment which science has barred from its domain ever since it began to rise rapidly in the seventeenth century. This is not to suggest that all scientists have always kept in mind this self-imposed limitation which freed scientific method from the encumbrance of heterogeneous considerations. That such a giant among scientists as Einstein frankly acknowledged that he could not derive ethical values from his scientific work,[2] did not prevent others from trying to deduce from science any and all ethics and values. As one may suspect, the ethics and values derived in such a fashion have a strange glitter. Something of it could be seen in that *New York Times* caption which greeted the publication of Monod's *Chance and Necessity*: 'French Nobel biologist says world based on chance leaves man free to choose his own ethical values'.[3] Most readers hardly took the view that the crux, possibly unintended, of that caption lay not with the words world, chance, ethical, and values. The pivotal word was man, a point which will require no proof for those aware of the role played by that word in the philosophical debates that led to Ockham's nominalism, and beyond it to empiricism and rationalism. The debate was about universals, usually exemplified by the notion of man and centred on the question whether there was a universal man, that is, human nature, or there were only individuals. Since the latter view has since been in steady ascendency, the caption could but carry the message that any and every man was free to choose, in a universe based on chance, his own ethical values. As in such a universe man is a mere accident, his ethical choices too have to be a mere matter of accident. For accidents, if they are truly such, one pays only penalty but bears no responsibility.

It is all too human to recoil from such a conclusion and to protest against such an inference from evolutionary views widely entertained since Darwin. But it is no less human to err and to do so in the most human way, which is to err by the shortness of one's memory. It is rarely remembered that no decade has passed since the publication of the *Origin of Species* without a prominent Darwinist spelling out bluntly the true message of the philosophy of natural selection. That message is the rule of the stronger, the rule of superman, the rule of the master-race. The following declaration of Haeckel is representative of utterances of which prominent Darwinists delivered themselves from time to time: 'The theory of selection teaches us that in human life, as in animal and plant life everywhere, and at all times, only a small and chosen minority can exist and flourish, while the enormous majority starve and perish miserably and more or less prematurely ... We may profoundly lament this tragical state of things, but we can neither controvert it nor alter it. "Many are called, but few are chosen." The selection, the picking out of these "chosen ones", is inevitably connected with the arrest and destruction of the remaining majority. ... At any rate, this principle of selection is nothing less than democratic, on the contrary, it is aristocratic in the strictest sense of the word'.[4]

Thus the enthusiasm for Darwinism of the advocates of the dictatorship of the proletariat and of a master race is all too understandable. Marx was quick to notice the usefulness of Darwinist theory for promoting class struggle,[5] and Hitler volubly echoed Darwinist views very popular among German military leaders prior to the First World War as a justification of their and his plans.[6] Preparation for class warfare, for trench warfare, for Blietzkrieg, and for wars of liberation have been repeatedly presented as the implementation of natural selection. Equally understandable is the readiness of those, in whose 'humanistic' philosophy Darwinism has a place of honour, to downplay its gruesome points. Haeckel's indebtedness to Darwin hardly came into view in a fairly recent book on the origins of National Socialism. Books on Bernard Shaw hardly ever contain a dishearteningly candid passage from the Preface of his play, *The Heartbreak House*, presented for the first time in the immediate wake of the First World War. There Shaw's victorious compatriots were reminded about Darwinism in particular and of science as the new religion in general: 'We taught Prussia this religion; and Prussia bettered our instruction so effectively that we presently found ourselves confronted with the necessity of destroying Prussia to prevent Prussia destroying us. And that has just ended in each destroying the other to an extent doubtfully reparable in our time'.[8]

For all that, Darwinism – as distinguished from the conviction fully shared by this author in the factual though still very imperfectly understood instrumentality of a species in the rise of another[9] – has a powerful appeal. Darwinism is a creed not only with scientists committed to document the all-purpose role of natural selection. It is a creed with masses of people who have at best a vague notion of the mechanism of evolution as proposed by Darwin, let alone as further complicated by his successors. Clearly, the appeal cannot be that of a scientific truth, but of a philosophical belief which is not difficult to identify. Darwinism is a belief in the meaninglessness of existence.

The celebration of meaninglessness

There are, of course, facets in Darwin's theory which should serve as a caveat for any perceptive and honest Darwinist to see Darwinism for what it has been taken time and again, namely, for a proof of the absence of anything permanent and well defined. Natural selection is powerless if there is no material substratum which is specific and stable, subject to stable and specific chemical and physical laws, and is therefore a meaningful entity. More often than not these facets are not seriously pondered even by those Darwinists who happen to mention them. And they are rarely if ever mentioned by the popularizers of Darwinism. This almost invariable slighting of facets strongly suggestive of meaning seems to be motivated by an urge to secure the meaninglessness of existence. The urge is not new. The logic-defying aphorisms of the Taoists, the escapism of Buddhists, the brazen commercialism of the Sophists, the stolid heroics of some Stoics, the standoffish smile of the Sceptics were in the pre-Christian world so many options for the meaninglessness of existence. In the post-Christian world the same option demonstrated anew its powerful appeal through the sudden popularity of existentialism. The workings of that appeal found a classic articulation in a statement by Aldous Huxley, who after insisting on the non-philosophical motivations underlying all systems, however philosophical, subjected his own philosophy as well to the same scrutiny:

> 'For myself as, no doubt, for most of my contemporaries, the philosophy of meaninglessness was essentially an instrument of liberation. The liberation we desired was simultaneously liberation from a certain political and economic system and liberation from a certain system of morality. We objected to the morality because it interfered with our sexual freedom; we objected to the political and economic system because it was unjust. The supporters of these systems claimed that in some way they embodied the meaning (a Christian meaning, they insisted) of the world. There was one admirably simple method of confuting these people and at the same time justifying ourselves in our political and erotic revolt: we could deny that the world had any meaning whatsoever'.[10]

Huxley's additional remark that the same tactic played an important role in the strategy of the chief spokesmen of the French Enlightenment is worth considering. Even more penetrating is his admission that, in his case too, the programme of meaninglessness could not be adhered to with consistency. The presumed unjustness of systems advocating meaning could be decried only if some 'just condition' was held high, be it the justice of radical anarchy standing for the alleged absence of any law and meaning. Revealingly, many an advocate of social anarchy ended up by taking it for a transitional stage to a rigid social order, a state of clear-cut meaning, desirable or not. As for Huxley's advocacy of sexual freedom, it pointed beyond its sweet anarchy to the *Brave New World*'s regimented system of procreation, an outcome as void of true sweetness as it is of anything truly new and brave.

It would, perhaps, be unrealistic to ascribe the most fundamental appeal of Darwinism to a motivation relating to sex. It would be certainly unrealistic, and especially in an age over which Freud had cast a long shadow, to take such a motivation too lightly. Theism, especially Christian theism, can be all too heavy a burden with its insistence on natural law which implies constraints and not in the least in matters relating to the use of sex. The desire to be liberated of such constraints can easily become an overpowering factor in making one adopt a *Weltanschauung* which will therefore gravitate in the direction of less law and therefore of decrease in meaning. The liberation achieved by casting one's lot with meaninglessness may be an exhilarating experience, though like other forms of exhilaration, hardly a permanent state. Were the state of exhilaration a permanent state, meaningless-ness would hardly be the proper word to describe it, because permanence implies coherence and therefore meaning.

A state or condition which by its permanence evokes coherence is certainly different from complete chaos, the only state void of all meaning. But just as an absolute chaos is unthinkable, so is complete meaninglessness. One can at most skirt some glaring evidences of meaning, a strategy obvious in customary presentations of Darwinism in which its facets indicative of meaning in a well-nigh metaphysical sense are methodically disregarded. Such a disregard of meaning should be more than enough for anyone interested in consistency to treat with contempt the meaninglessness which Darwinism is meant to support. Those with such interest have, however, always been in a minority. The wider public does not indeed notice today the somersaults in logic in current speculations on extraterrestrial intelligence (or ETI for short), this ultimate extension of Darwinism and an utterly self-defeating exercise in wishful thinking. A strong phrase like this, equivalent to a *lèse majesté*, carries the risk of being taken out of context and held high against its author as if he had never stated his belief in the fact of evolution, that is, in the instrumentality, however poorly understood, of a species in the rise of another. Yet, the need for such a strong phrase is overriding as nothing else can even be modestly effective in drawing attention to that strong-headed exercise in methodical meaninglessness which sets the tone of almost all that is being published and said nowadays on extraterrestrial intelligence, or ETI. Fashionable speculations on ETI, nay brave assertions of its factuality, are so many rigged testimonials in a courtroom set up to banish God from the realm of the living, to pre-empt purpose of all its significance, and to assure the rule of meaninglessness.

Extraterrestrial intelligence

Only a few advocates of ETI speak their minds openly on what they consider to be the supreme fallout from an eventual detection of ETI. Such an outcome, so they believe, would be the final rebuttal of supernatural revelation, the ultimate unveiling as sheer myths of such tenets as Fall, Incarnation, and Resurrection.[11] But all those, who see in that outcome the proof that the mind is a mere epiphenomenon of biochemical diversification, reveal enough of their belief that the universe is not something which owes its existence to the Intellect and

Will of the One WHO IS, that is, existence itself. Actually, what should be branded as sheer myth is the claim that the reduction of the mind to complex biochemical processes has been scientifically demonstrated. Any serious monograph on brain research reveals enough of the perplexities of even such leaders in the field who are not at all sympathetic to dualism, that is, to the ultimate irreducibility of mind to brain.[12] That Sherrington, universally recognized as the foremost student of the brain in this century, was an emphatic voice on behalf of dualism, is something not to be written off easily,[13] although hardly ever put in writing by authors of shallow though highly acclaimed science popularizations, in which, to mention only one example, Broca's brain is held high[14] but not the brains of Sherrington and others. Such is, however, the procedure of courtrooms where the alleged objectivity of a scientifically coated view is sheer virtue and its criticism on the ground of thorough science is sinistrous vice. The same court has no room for the evidence of countless living brains, the brain of aborigines, whom Darwin and others complacently presented as the 'missing link' between men and apes, capable of communicating only with grunts and inarticulate cries. Alfred R. Wallace, the co-proponent of the theory of evolution by natural selection, who observed them for years at close range, had a very different impression which had a devastating implication for the theory. The brain of primitive man, Wallace wrote, which is 'but very little inferior to that of the average member of our learned societies', was 'an organ . . . developed far beyond the needs of its possessor'.[15] As it turned out, the development of the human brain not only took place in advance of human needs, but also with an incredible rapidity, within practically one instant on the geological scale, as was put by a leader in paleoneurology.[16] Against such conclusions Darwinians are reduced either to an authoritative No!,

Darwin's 'scientific' response to Wallace's paper, or to a lame silence.

The evolution of the brain as the organ of thinking is but the top of that iceberg, Darwinism, which must leave cold the sober investigator of any layer of it. Its successive layers corresponding to successive geological ages should be abundant in all sorts of transitional forms. Darwin's own admission, that the failure of geological research to yield the infinitely many fine gradations between past and present species as required by the theory is 'the most obvious and serious objection which can be urged against it',[17] remains as relevant as ever. What most effectively gives away Darwinism is the almost mystical faith voiced by its supporters in facing up to the absence of evidence and even to the contrary evidence. T. H. Huxley, who brushed off Kelvin's weighty objections with the remark, 'We are Gallios who care for none of these things',[18] set a pattern no less than did Haeckel who thundered: 'The history of evolution is the heavy artillery in the struggle for truth'.[19] Only on a rare occasion there comes along a Darwinist able to present his weighty generalizations with a light touch. Some of these must indeed be taken lightly, if the learned investigator of the rise of vertebrates, half a billion years ago, was entitled to say: 'There is no direct evidence that any of the suggested events or changes ever took place. . . . In a sense this account is science fiction. . . . Unfortunately, there is little hope that the morphological argument can find verification in the discovery of actual fossil animals, which would . . . give intellectual respectability to our procedure. . . . Proof may be for ever unobtainable, and it may not matter, for here is such stuff as dreams are made on'.[20]

Darwinists – who in no way ought to be confused with those who believe in evolution but not in the all-purpose character of this or that mechanism, such as natural selection, proposed as its full explanation – have indeed been great dreamers, a quality needed in all creative enterprise, including science. In dreaming about a 'warm little pond with all sorts of ammonia and phosphoric salts, light, heat, electricity, etc, present'[21] as the place of spontaneous generation, Darwin prophetically anticipated by a hundred years scientific speculation and even experiments. But was not Darwin equally prophetic in stating that even if such a pond – replaced today by the primeval ocean exposed to high-intensity electric discharge – could be adequately described, the origin of life would still elude us? The distance from amino acids obtained in experiments simulating primeval conditions to really living cells is no less immense than it was in Darwin's time. Indeed, the startling progress made in biochemistry made possible very concrete estimates as to the enormous improbability that a living cell, or even any of its major constituents, should arise through a chance process. The more is known about the evolutionary history of the earth, the more narrow becomes that period in which life could arise even if theories of the origin of life were considerably more reliable than they are.

Staggering figures of the various aspects of the improbability of spontaneous generation have been provided ever since Darwin at regular intervals though with little effect. Such a result must have its explanation in factors which only a thorough psychoanalysis would ever fathom. One seems to be in the presence of deep-seated mystical urges when one considers, for instance, the runaway popularity of estimates of the number of planets with civilizations in the Milky Way, and even in the entire universe. The formula $N = \int R_{\star} f_p n_e f_l f_i f_c L$,[22] – first construed by F. D. Drake and seized upon by Sagan with crusading vehemence – which allegedly yields that number for a galaxy, proved to be a powerful stimulant for scientific amateurs and scientific illiterates given already to the belief in

ETI. Little impact, on the contrary, was made by W. H. McCrea's evaluation of that formula. His testimony is all the more valuable because he is the author of one of the most brilliant and yet basically inadequate variations of such theories which suggest that there is a planetary system around a good portion of stars, whose number, in the Milky Way alone, is millions of billions. Sagan's formula is useless, McCrea remarked, because it is without foundation since there is no reasonably good theory of how our solar system, the only known system of planets, has evolved.[23]

That such a remark fails to make impact can be understood from a brief survey of the history of theories of the origin of planetary systems.[24] The history begins with Descartes who is also the starting point of theories in which a star generates around itself a system of planets. The other main type of theories is the collisional type in which a planetary system is a mere accident. Although the former type of theories is no less beset with difficulties than the latter, it has always enjoyed far greater popularity of which the renown of the so-called Kant-Laplace theory or nebular hypothesis is a chief example. It now reigns again in vastly revamped forms, after a short period in the early decades of this century when the collisional theory saw a brief revival. Almost invariably, the non-collisional theories contain a hint, and at times even a boast, as to the satisfaction felt by their authors that the presence of intelligent beings everywhere in the universe is made thereby plausible if not simply assured.

What is actually assured, even if one grants as facts two cardinal tenets in the Darwinian creed, the spontaneous origin of life and the spontaneous emergence of mind? On the basis of that creed, an earthling should feel anything but assurance if planetary systems and life-bearing planets are a regular occurrence in the vast stellar realms. If Darwinism is true, evolution in most other solar systems must have followed lines very different from the one it did on the earth.[25] Actually, according to Darwinism, evolution on our earth would be considerably different if it started anew. As to other solar systems, evolution there may have far surpassed the stage at which it now finds itself on the earth. On a large number of planets, perhaps on some planets within a few dozen or hundred light years from us, there should then exist, were Darwinism true, not only men with superior intellect and technology, but mammals of inconceivable shapes as superior to man as man is to apes and perhaps to much lower animals. In other words, those other civilizations will have for their builders most diverse species of intelligent beings, who, according to Darwinism, cannot be expected to display more 'humanness' toward us than what is shown by us toward our domesticated animals, many of which we keep for convenient protein reservoirs. Indeed, we earthlings would have for some time been reduced to the status of cosmic domestication and for a reason which has very recently gained quite a recognition and which also reveals Drake's formula as a most potent mystery mongering. The formula is also a transparent fallacy. Whatever the merit of envisioning the spontaneous emergence of higher civilisations around many a star, it is not a question of visionary view but of sober observation that the great majority of stars is older than the sun. Consequently, supercivilisations around them must have been thriving for long enough a time that their spaceships should have already explored our planetary system[26] and landed on our earth with disastrous consequences for us. This inference can be evaded within the Darwinian perspective only by a recourse to the vastness of the universe, but in this case our own search for ETI must also be considered an enterprise far less promising than looking for a needle in a haystack.

In rebuttal of such considerations the Darwinist cannot argue that those superintelligent species are endowed with the same type of intelligence we boast of, superior as theirs may be to ours in many or perhaps most cases. Darwinism has as little room for such an argument as it does for a recognition of something stable and universal in man's nature. Intelligence as a capability which admits various degrees within essentially one type, so as to make possible rational communication among them (to say nothing of a common recognition of universally binding moral laws), can be argued consistently only within a metaphysical realism, but never within a philosophical discourse which rests, directly or indirectly, on nominalism. While the latter radically bars the mind's road to God, the former naturally culminates in a theism of which the doctrine of creation is an integral part. Only such a theism, for which finite intellects are the highermost image of God, the infinite intellect, contains logically the assurance that all finite intellects in the universe should think and feel along essentially the same lines. A theist, and only a theist, can therefore look forward with confidence to an eventual encounter with intellects from other solar systems, because the encounter will be between two sides, both of which, though possibly in a different degree, will know something of a universal brotherhood based on common dependence on the Creator. This is why a theist will not side with those who urge, as did some Nobel-laureates (undoubtedly with that cosmic protein reservoir in mind which is the real physiognomy of cosmic connection within Darwinian perspective), that the sending of messages into outer space be discontinued lest we reveal our existence to extraterrestrial beings.[27] But the theist cannot side with those either who urge the financing of prohibitively costly projects for the detection of ETI, projects based on the philosophy (if it can be called such) that intellects are a mere epiphenomenon of biochemical diversification and therefore thrive in every nook and cranny of the universe. Only a theist, for whom intellects are a special creation of God, can look at the question of ETI as a truly open question which cannot be prejudged scientifically. Clearly, no one can prescribe to God to create intellects everywhere or to limit his power to do so.

A theist can only look with pity at Darwinist advocates of ETI. Nothing short of pitiful are the somersaults in logic with which they try to usher through the back door that universality of intellect – a glaring though very necessary departure on their part from their stated belief in universal meaninglessness – which they had already banished through the front door by their initial presuppositions. Those somersaults are so many rapes of logic which hardly arouse the indignation of an age which finds itself, through the steady and systematic lowering of ethics into the meaninglessness of sheer patterns of behaviour, more and more unable to denounce rapes of much more tangible kind. A recall of that systematic rape of one's memory which advocates of ETI are almost obliged to commit in order to save face will therefore be all the more telling. Its feasibility is based on the vast amount that can safely be assumed to have been erased from public memory in less than two decades. Only a few historians would recall today that in the early 1960s a blue-ribbon panel of scientists assured, on behalf of the National Academy of Sciences, the government of the United States that low-level organisms, nay plants similar to moss, are most likely thriving on Mars,[28] an assurance which led to the implementation of a multibillion dollar project designed to detect the existence of life there. Again, very few recall today that the pictures of Mars sent back by the space probes Mariner 6 and 7 were the kind of 'stunning crop of surprises' – an editorial phrase in the *New York Times*[29] – which poured cold water on sanguine expectations. The pictures showed Mars to be pockmarked with craters no less than is the moon, hardly the expected environment for life at however modest level. The final phase of the project was no less sobering in its outcome. A most sensitive multiple analysis of Mars' soil by an instrument landed on it in 1976 failed to detect any trace of life, present or past. The result fully deserved the biting remark of a scientist in charge of that analysis: 'There's every sign of life, except death'.[30]

Memory can fade even in a short year or two. Sagan, one of the chief instigators of the project and undoubtedly the high-priest of advocates of ETI, was installed in 1978 by the President of the United States in Blair House to spread the Gospel of ETI from that exalted and plush federal forum. Sagan claims, of course, that the result of the Mars-soil analysis is merely 'inconclusive'. Yet, even if such were the case, the result is in shocking discrepancy with the prospects held high less than twenty years ago. Of that discrepancy Sagan does not remind his enthusiastic audiences and his avid readers. It is not difficult to guess what he will tell them after Pioneer II had taken a close look at Titan, Saturn's largest moon, dubbed by Project Scientist John Wolfe as 'sort of the [exo-] biologist's last hope'.[31] No doubt Sagan will perform his usual verbal footwork in which a 'titanic' frustration appears in the end a 'qualified' success.

Empty plenitude

Those who are not Sagan's followers and are not scientists either will perhaps wonder how the same data can give rise to widely differing evaluations. The explanation lies in the subjectivity inherent in scientific work in spite of its reputation of being the paragon of objectivity. Very revealing in this respect is the history of speculation on ETI of which a memorable phase stretches across the second half of the eighteenth century, a period of almost a hundred years after the publication of Newton's *Principia.* A curious circumstance, because there was enough in the *Principia* and in the physics and astronomy built on it during the next two or three generations to make gravely suspect brave assertions of the existence of intelligent beings on other members of the planetary system. Yet, a Herschel confidently peopled even the sun![32] Shortly before, Lambert, another scientist of considerable stature, emphatically asserted the presence of intelligent beings on all comets. Reference to asbestos disposed, in his view, of the problem of resistance to sizzling heat on those comets whose perihelion was so close to the sun as to be well within the orbit of Mercury.[33] If a Herschel and a Lambert and a great many other contemporary scientists with similar views reasoned in such an astonishing manner, the reasoning of Kant, a mere amateur in science, who specified not only the physical but also the intellectual and moral characteristics of denizens on all the planets, will appear less astonishing.[34] At any rate, the eighteenth-century infatuation with the presumed presence of intelligent beings on almost every celestial body will appear astonishing only if one overlooks the overpowering impact which a commonly shared philosophical presupposition can have on thinking about matters scientific. In this case the philosophical presupposition was the principle of plenitude, according to which the created realm has to reflect in every possible way the plenitude of God. Since God's life is intellect, life and intellect must be everywhere in the universe.

This necessitarian and *a priori* character of the principle of plenitude should make it clear that its origin is not Christian, whatever its injudicious endorsement by many a Christian thinker. It was given its most telling articulation by those who were in turn the spokesmen of animistic pantheism or of a rationalism which could not, even when it wanted to, be theistic. A classic representative of the former was Giordano Bruno, who in some circles, woefully behind in historical scholarship, is still a martyr of science.[35] As a matter of fact, Bruno's pantheistic denial of all tenets of Christian faith and his gross obscurantism were so glaring in his own times that he had no influence until the pantheistic German idealists resurrected him into spurious glory. No less instructive is Leibniz as a representative of that rationalism. His thinking embodies a dichotomy observable in not a few Christian thinkers: a deep religiosity coupled with a shallow philosophical notion given a central role. The pivotal claim of Leibniz, that creation had to issue in the best of all possible worlds, could but deprive of its depth his insistance on the radical contingency of the universe.

The dichotomy will surprise no one moderately aware of the extent to which Christian thinkers can fall prey to the prevailing fashions of thought and of the extent to which they are prone to fall back at all times on that most pagan of all ways of thinking, which prompts one consciously or subconsciously to eliminate the vista of contingency. Any attempt to understand the principle of plenitude, or the present-day infatuation with ETI, which ignores that resolve, will result not in a fathoming of the depths, cosmic or philosophical, but in a splashing on the surface. Actually, more than mere awareness of that resolve is needed to take a proper measure of it. Lovejoy, the foremost historian of the principle of plenitude, was certainly aware of that resolve. What he was not aware of, or rather not sympathetic to, was that meaning of contingency which only a genuine theist is able to grasp. That contingency is a far cry from the one formulated by Spinoza whose definition of contingency was quoted by Lovejoy as he closed his classic monograph, first published in 1936, with a brief reflection on the 'emptiness' of the universe as revealed by modern astronomy.[36]

In the 1930s the 'emptiness' of the universe as far as life was concerned was strongly intimated by two developments in astronomy. One was the prevailing belief that the only reasonable explanation of the origin of the planetary system lay in a near-collisional theory such as was mainly articulated and popularized by Jeans. The other was the disrepute into which wishful thinking about planets as possible abodes of life had fallen. Nobody, for instance, took seriously earlier reports about the existence of 'canals' on Mars. Also there was a noticeable increase in the reliability of estimates of the variation of the surface temperature of various planets, hardly a favourable result to encourage viewing them as abodes of life. The popularity of Darwinism was also at a low ebb at that time. Today the 'emptiness' in question is far more striking in spite of a reversal in thinking about the origin of planetary systems which are now viewed again as regular accompaniments of a large portion of stars. That the reversal was motivated in part at least by the desire to secure planetary systems in very large numbers should intimate something of the wishful thinking lurking behind the theories now in vogue about their origins.

Thinking when done by scientists is not necessarily the source of truth, scientific or otherwise. More likely than not, it is suggestive of presuppositions whose truth rests with the scientist's philosophy and not with his science. How misleading that thinking can be is well illustrated by a detail which has been heavily relied upon by advocates of ETI for whom observational evidence of at least one planet around a nearby star would represent a clinching proof about the presence of planets around any star. They certainly took in this sense the observation by Van de Kamp of a waviness in the path of Barnard's star, which of course may indicate the presence around it of a planet as heavy as Jupiter and perhaps of another as large as Saturn. While inference from that waviness to the existence of a planet is perfectly reasonable, the complacency which precluded the repetition of Van de Kamp's observations should seem rather unreasonable. Tellingly, when finally this was done, the results were far less impressive and even a basic fault was found with the mounting of van de Kamp's telescope.[37]

Failure to observe a planetary system other than ours and to devise a theory of the origin of our planetary system free of serious difficulties is, clearly, a failure to provide a reasonable basis for inferring the existence of other systems and to estimate their number. This, however, must not be taken as a proof of the uniqueness of the solar system. Nothing is more self-defeating in matters scientific than to declare that persistent failure to discover or to explain something means that it will remain forever beyond our grasp, intellectual or observational. The entire history of science is a witness to the truth of the principle that the unknown of yesterday becomes the novelty of today and the commonplace of tomorrow. It is also true, however, that science has revealed so far a vast array of evidence which should force the unprejudiced mind to look with wonderment at our earth as an abode of intelligent life. The extreme fragility of the earth's crust is not the only factor which imposes enormous restrictions on the evolution of life. A similar role is played by the quantity and composition of the atmosphere and of the oceans. As for the latter, a minor change in their salinity could be destructive of much of marine life. These physical characteristics of the earth, setting narrow limits to the evolution of life on it, are in turn the function of its density, chemical composition, distance from the sun, and of the mass and size of the sun. Last but not least, there is the problem of the earth's position among the planets, a very delicate position indeed. The planetary system is such a finely tuned mechanism that if the mass of Jupiter were to be increased by one percent,[38] the planetary system would no longer have that stability which is a precondition of life on earth.

The existence of intelligent beings on the earth should become no less an object of wonderment when a look is taken at the evolutionary tree. It is a tree which obviously could have grown in many different ways. Otherwise that tree would not show an immense preponderance of branches which led nowhere over those branches that can be considered as leading to the evolution of mammals. As for the many branches that stand between mammals and man only those can so far be seen clearly that certainly did not lead to him. To think that evolution would run the same course again, would be, as was already noted, in defiance of all basic postulates of Darwinism. Even more un-Darwinian is the postulate, or rather very Darwinian because of the wishful thinking embodied in it, that evolution on other planetary systems would issue in beings similar, however remotely, to man. That not even the Darwinian exobiologist has the privilege of eating his cake and having it too is the gist of the eloquent passage whose author, a judicious Darwinist, unwittingly claimed that privilege as he swept cosmic spaces clear of men, but not of minds similar to man's:

> 'Lights come and go in the night sky. Men, troubled at last by the things they build, may toss in their sleep and dream bad dreams, or lie awake while the meteors whisper greenly overhead. But nowhere in all space or on a thousand worlds will there be men to share our loneliness. There may be wisdom; there may be power; somewhere across space great instruments, handled by strange, manipulative organs, may stare vainly at our floating cloud

> wrack, their owners yearning as we yearn. Nevertheless, in the nature of life and in the principles of evolution we have had our answer. Of men elsewhere, and beyond, there will be none forever'.[39]

The luck of science

Such a conclusion will appear far less speculative if one considers the much more concrete question of what man would be able to do elsewhere in the planetary system. He would certainly not speculate on ETI on the basis of a highly developed astronomy. On any other planet he would have been barred, or at least greatly impeded, from acquiring a significantly quantitative knowledge about the solar system and the system of stars.[40] In the absence of such a knowledge he would not know Newtonian dynamics, let alone modern physics, and therefore nothing of that biophysics and biochemistry by which speculators on ETI set so great a store. Man's indeed is a lucky place in the solar system to cultivate astronomy. Only from the earth, and not from any of the other planets, have the moon and the sun the same apparent size which makes possible total eclipse, a phenomenon of crucial importance for the development of astronomy. The same apparent size of the moon and of the sun made possible many of those attainments which are the basis of Ptolemaic astronomy, without which Copernican astronomy and Newtonian physics would be inconceivable. The two satellites of Mars are much too small to be seen with the naked eye by a man placed on its surface. A Newton there may have had an apple perhaps, but certainly no moon to verify the universality of the law of fall. As to the satellites of Jupiter and Saturn, several of them appear much bigger than the moon and the sun appear to us, and from each of those planets the sun is hardly more than a bright star and not that imposing though not overpowering fiery body which could prompt, as it did in our case, the recognition, however belated, of the heliocentric ordering of planets.

In addition to the earth's position in the solar system, the inclination of its axis too is a lucky circumstance for the development of astronomy. That axis has been pointing (within historically recorded times) toward a star sufficiently bright to make possible the pre-telescopic observation of the precession of the equinoxes, a pivotal feat indeed. Astronomy would have hardly fared well were the larger part of dry land in the Southern Hemisphere from where no similar polar star is visible. One should perhaps be less ready to deride eighteenth-century authors of natural theology, such as Derham, who saw in the four seasons as caused by the inclination of the earth's axis a finger of Providence at work.

The earth's position in the Milky Way is no less favourable for astronomy. Were the sun near the centre of the Milky Way, the latter would not be visible even with telescopes owing to the thick interstellar clouds there. Furthermore, the much stronger cosmic radiation near that centre would make life impossible. But even in the outer arm of the Milky Way, where the sun is actually located, a small cosmic cloud might just have forever blocked from view the famous Crab nebula, an observational goldmine for astronomy. Also, were the earth in a slightly different position with respect to the moon, the latter would never occult the brightest of all quasars so as to permit its proper study. Would it now be reasonable to assume that in other planetary systems there would be an earth so carefully positioned as to provide its inhabitants with the same extraordinary circumstances for developing astronomy (a most crucial component in developing science) as are available on our earth?

A similar reappraisal of man's geological and biological environment would in all likelihood be rich in intimations as to how peculiar is what makes man, namely, his cultural productivity. Indeed the whole history of man is a series of startling peculiarities. The farther that history is being pushed back into the past – now perhaps as long as two or three million years – the shorter becomes the time needed within the perspectives of Darwinism for the evolution of *homo faber* from still hypothetical monkeys. At the other end of that history the peculiarity of it should seem to be simply overwhelming. Is not there something enormously peculiar in those few thousand years, which witnessed man's cultural rise, and in those last few hundred years, which witnessed the rise of science? Why is it, to mention only one example, that the Chinese, who were for many centuries in possession of gunpowder, printing, and compass (three inventions in which Francis Bacon saw the origin of science) failed to come up with a self-sustaining science, the only kind of science worthy of that name? Why is it that science suffered a stillbirth in all ancient cultures, none of which lacked in talents and in spectacular technological feats, among them the pyramids? The Greeks' failure to raise science to a self-sustaining level is, of course, the most tantalizing problem both for those who see science as the necessary outcome of socioeconomical conditions and also for those who see in science the creativity of the mind, though not a created mind at its very best.

The creativity of the mind when severed from its ontological foundation which is the created mind is an artistic cliché. It is no match even for the feat of performing elementary arithmetic, let alone for that most puzzling performance in the long cultural history of man, namely, the apparently much delayed and explosively sudden rise of a science which feeds on its own success. Since the time of Bacon each century witnessed a wave of explanations about that sudden rise. Bacon could not depart from the idea, required by his empiricism, of slow accumulation, the very opposite of what was to be explained. In Bacon's century it has also become fashionable to speak of the genius of the times. Such self-appraisal was not altogether inappropriate for times that centuries later were aptly labelled the century of genius. But the notion of genius could not be handled either within Baconian empiricism or within Cartesian rationalism. The former failed to do justice even to a modest mind, the latter immodestly derived all from the mind. Jesuit missionaries in China, who made much of the superiority of European science, also failed to say anything meaningful about the genius of Europe responsible for that superiority. Genius, as used in the seventeenth century, was reflective of acumen infected with smugness. This is why the seventeenth-century popularity of the phrase, 'we see farther because we are sitting on the shoulders of giants', a phrase coined by the twelfth-century logician, John Salisbury, was mostly a lip service to intellectual humility, the hallmark of true genius.[41] The true historical significance of the phrase remained hidden in a century which failed to esteem that its finest technological monuments, the gothic cathedrals and the mechanical clocks operating with the aid of a double feedback mechanism, were of medieval origin. Much less was it recalled that Descartes and Bacon merely echoed medieval schoolmen by celebrating man as the master and possessor of nature. Long before Descartes and Bacon the systematic conquest of nature had been under way and under the impact of a widely shared belief in the biblical account of creation which put man on a unique pedestal only to be turned into a Tower of Babel by Cartesian rationalism and Baconian empiricism.

The eighteenth-century efforts to account for the sudden rise of science were vitiated by the smugness with which a rationalist (Cartesian) Enlightenment applauded Bacon's empiricism. In accordance with the latter, Turgot and Condorcet defined genius as the result of the accumulation of many mediocre talents.[42] Great painters, Turgot argued, for example, could be found only in countries where many insignificant artists were selling their wares. The argument certainly was not applicable to China, the nation with the greatest number of painters, craftsmen, and accountants. Yet, in a display of the blindness generated by a smug Enlightenment, the origin of science was placed by Bailly in a mythical prediluvian kingdom, occupying the land which is presently known as Mongolia. Little did he suspect that the origins of modern science were rather in his home town, Paris. But medieval Paris was not to be preserved, let alone to be studied, by those whose great ambition was to turn Paris into a City of Light.

Bacon and his contemporaries, convinced Christians as they were for the most part, saw something of the role of the faith in creation in the rise of science. Yet neither the Protestant nor the Catholic side saw anything significant in the fact that this faith first became a widely shared cultural matrix in the medieval centuries. These, tellingly enough, received the label Middle Ages in the seventeenth century, a label which represented not much improvement on the epithet, dark ages, in vogue since Renaissance times. Sensitivity to the role which a widely shared belief in creation could play was even weaker among the Encyclopedists, many of whom tried, though in vain, to establish a rational or rather deistic middle ground between Christianity and materialism. It was with no great conviction that they recognized their debt to medieval schoolmen for the rational articulation of the notions of Supreme Being, creation, immortality of the soul, and the like, which were needed by Deism. References to Creator and creation were absent in nineteenth-century theories on the origin of science. Their authors, Comte, Spencer, Mach, and others, are best remembered for their philosophies in which there was no room for the Creator, nor for that matter, for creative science. This kind of science could not be in view in decades when it was widely believed that science, especially in its fundamental form, physics, had given all the explanations of which it was capable and that hardly anything was left to be explained.

While the twentieth-century amply revealed unsuspected depths of scientific creativity, it was not these developments that produced a break with already hardened, mostly positivist, clichés on the origin of science. The break was largely effected by two monumental investigations. Of the two, the ten massive volumes of *Le système du monde* by Pierre Duhem, who knew by heart the *Imitation of Christ* and Pascal's *Pensées*, could not be to the liking of a century eager to affix on itself the label 'post-Christian'. Still, since the publication of the *Système du monde*[43] it has at least become unscholarly to ignore entirely the medieval origins of modern science, although it is still scholarly to ignore completely the relation of that origin to the belief, widely shared during the Middle Ages, in *the* origin, that is, creation. That half a century after Duhem, the Middle Ages and Christianity were decried by Popper as the factors that suppressed science,[44] tells something of the true nature of the thinking which generates a 'logic' of scientific discovery according to which no logic is to be found there. Within the perspective of such logic nothing worthwhile will appear in the philosophy of the two millennia separating the pre-Socratics from Descartes. As for medieval philosophy, to take it for an effort to count angels on a pinhead is as scholarly as to take Plato for a blind sage trying to see his own shadow in a dark cave.[45]

The other work in question, Joseph Needham's multivolume *Science and Civilisation in China*, is, of course, on the lips and pen of everybody in an age applauding Marxism and focusing on China as the geopolitical giant of the not-too-distant future. Needham's study bears witness everywhere to his professed Marxism, which is, however, not rigidly Party-line. More regular Marxists found indeed unpalatable one of Needham's major conclusions which he reached after many years of systematic effort to prove that scientific and technological history of China bears out Marxist theory.[46] The conclusion states nothing less than that the failure of science in ancient China is a failure of the ancient Chinese mind on a point of theology, the doctrine of creation. According to Needham, even if belief in that doctrine had been alive in the very earliest phase of Chinese history, it soon petered out with a fateful result for science in China: 'It was not that there was no order in Nature for the Chinese, but rather that it was not an order ordained by a rational personal being, and hence there was no conviction that rational personal beings would be able to spell out in their lesser earthly languages the divine code of laws which he had decreed aforetime'.[47]

A conviction to recover

The key word in the foregoing passage is conviction, standing for trust in the rationality of created nature as well as in the powers of created minds. Such a trust was clearly present in the Christian West, the birthplace of science, where the Christian dogmas of creation and final consummation implied belief in a linear progress, however mysterious. The absence of that trust in other cultural contexts ought therefore to appear most significant. The despondency, which sets the tone of all major ancient cultures, is indeed of a piece with the defeat science suffered in each of them. The world view of all those cultures was based on the doctrine of eternal returns, cosmic and human, which invariably generated the sense of being trapped in a treadmill. The exclamation of King Brihadrata, 'In the cycle of existence I am like a frog in a waterless well',[48] is only one of the many similar utterances prompted by a defeatist outlook in which the universe could only appear a trap from which there was no escape.

It was only within Christianity as a social matrix, that there arose the broadly shared conviction that existence, cosmic and human, was not a trap, precisely because the universe could be viewed as a home once Creator and creation were in full view. Such a view is a commitment to taking for real and meaningful some beacons of light cutting through a vast darkness, the darkness of man's very limited knowledge and of his almost unlimited tragedy of suffering. Once severed from that view, that is, from faith in the Creator, references to the universe as a home become well-meaning rhetoric. Of course, science is now in possession of such a vast interconnection of data, laws, and instruments as to continue its progress even if no attention is paid any longer to that faith which played an indispensable role in its rise. The inner wealth of science can even withstand, without any recourse to that faith, the absurd philosophies of science which imply the denial of the objective reality, of the coherence, and of the uniqueness of the universe. Modern cosmology too will prove itself immune to the disease of solipsism and anthropocentrism which trapped some of the best cosmologists of our times. But without a reference to the doctrine of creation, which implies both the rationality and contingency of the universe and the matching rationality and contingency of *homo faber,* there will be no consistent explanation of the scientific method, a most delicate interplay of experimentation and reasoning, nor will there be a consistent correlation of cosmological theories, however powerful, with the cosmos, the coherence of all real contingent things.

More importantly, without a widespread recovery of belief in the doctrine of creation, there will be no assurance that science will not prove itself that tree of knowledge of which the one in the Garden of Eden is the primeval symbol. Since 'myth' is no longer synonymous with sheer fiction, the truth-content of that biblical myth should seem enormous to anyone mindful of the disastrous consequences which can follow from the indiscriminate tasting of the fruit of that tree which is scientific knowledge. In a far more tangible manner than the first man could dream of, the fruit of knowledge can in this age of science be a prompting to assume the role of God. The playing of God by dictators, past and present, is mere child's play in comparison to what can be done today by even a single scientist. Recent work on the genes of E. coli made it all too clear that careless handling of a single test tube may lead to a global disaster far more destructive than the accidental detonation of a hydrogen bomb can be.

If the episode of 'recombinant DNA' will prompt a return from sheer utilitarianism to moral perspectives only in the measure advocated by Rousseau, the prospects will not become less gloomy. A new social contract, as radically severed from belief in creation and Creator as the one advocated by Rousseau, will only produce another Utopia about noble savages. It is undoubtedly most ennobling to think that every person would 'give up a measure of his freedom in return for the protection of freedoms for all, so as to maintain a desirable human condition'.[49] Indeed, human nature may not be so corrupt as to prevent many from considering giving up, under the threat of global destruction, a measure of their freedom of action. But in order for this consideration to be translated into action, something specific ought to be agreed upon about that measure and that condition. Specifics will be lost in sheer relativism and subjectivism unless agreement is reached about the generic question of the condition and freedom of man. If the history of modern philosophy, avidly in search of that agreement but no less resolved to skirt the doctrine of creation, is an indication, agreement will not be forthcoming in an age which boasts of its being post-Christian, a condition worse in some respects, than being pre-Christian.

The cure of modern philosophy will perhaps come from a reflection on its persistent malaise. Modern philosophy is caught, more than was the case with ancient Greek philosophy, in a hapless oscillation between two extremes: the mirage of absolute certainty and utter scepticism. Philosophy transcended this predicament only when during the Middle Ages a widespread adherence to the doctrine of creation kept philosophy on the ground of metaphysical realism, the only safe ground between the abysses of an absolute certainty bordering on tautology and a no-certainty-at-all provoking despair. Not until that ground is recovered, will there be an unequivocal surrender to the reality of the cosmos, the indispensable stepping stone for a rational recognition of the existence of the Creator. Such is the only solid basis for looking at existence not as a global trap and for mustering the resolve to view the earth a genuine home. But just as that ground was not gained in the Middle Ages, the age of faith, without the guidance of faith, it will not be regained without first recovering that faith. In a very crucial sense, one must first say Creator in order to say Cosmos.

NOTES

Chapter One

1. A. S. Eddington, *The Nature of the Physical World* (Cambridge: University Press, 1929), p. 338.
2. A. S. Eddington, *Space, Time and Gravitation: An Outline of the General Theory of Relativity* (Cambridge: University Press, 1920), p. 201.
3. J. H. Jeans, *The Mysterious Universe* (New York: Macmillan, 1930), pp. 140 and 144.
4. See C. Kahn and F. Kahn, 'Letters from Einstein to de Sitter on the Nature of the Universe', *Nature* 257 (1975), p. 451.
5. G. Lemaître, *L'Hypothèse de l'atome primitif: Essai de cosmogonie* (Neuchâtel: Editions du Griffon, 1946), p. 90.
6. Aquinas' view, which he set forth in varying details on six different occasions (including his two *Summae*), is all the more noteworthy as it countered and reversed an already hallowed consensus among Christian theologians from Augustine to Bonaventure that the temporarality of the universe can be demonstrated by reason. For the historical background, texts and documentation, see *St. Thomas Aquinas, Siger of Brabant, St. Bonaventure: On the Eternity of the World (De Aeternitate Mundi)*, translated from the Latin with an Introduction by C. Vollert, L. H. Kendzierski, and P. M. Byrne (Milwaukee, Wisconsin: Marquette University Press, 1964).
7. A. S. Eddington, *New Pathways in Physical Science* (Cambridge: University Press, 1935), p. 60.
8. *Ibid.*, pp. 58–9.
9. A. S. Eddington, *The Expanding Universe* (Cambridge: University Press, 1933), p. 56.
10. *The Nature of the Physical World*, p. 85.
11. For the papers presented by the panelists, see 'The Evolution of the Universe', in *Nature* 128 (1931), pp. 697–722, or in *British Association for the Advancement of Science. Report of the Centenary Meeting, London, Sept. 23–30, 1931* (London: Office of the British Association, 1932), pp. 573–610.
12. *Ibid.*, pp. 602–4.
13. *Ibid.*, p. 604. Smuts saw a fresh support for the philosophy of emergence in the new quantum physics, according to which, as he put it, 'the roots of life and mind lie embedded deep down in the ultimate structure of this universe, and they are not mere singular apparitions of unaccountable

character, arising accidentally in the later phases of evolution' (p. 605).

14. *Ibid.*, p. 597.
15. *Ibid.*, pp. 578–9. As to the ultimate fate of the universe, 'a rapid spreading of ripples in a still more rapidly-spreading space', Jeans found comfort in the thought that 'when the worst comes to the worst, we shall none of us be there to see' (p. 579).
16. *Ibid.*, p. 608.
17. *Ibid.*, p. 607.
18. *Ibid.*, p. 588. This limitation of time is another aspect of the specific rate of expansion necessary for the production of the actual chemical elements and their relative abundances, that is, for the emergence of the actual universe as we observe it, a point to be discussed in the next chapter.
19. *Ibid.*, p. 582.
20. *Ibid.*, p. 586. If such was the case, radical positivists, like P. W. Bridgman, were right in pleading for the limitation of cosmology to a few million light years, about the distance of the Andromeda nebula.
21. *Ibid.*, p. 584.
22. *Ibid.*, p. 584. A memorable expression of that wish 'not to do so' is the remark in Einstein's 'Autobiographical Notes' which deplores 'the dogmatic rigidity' that prevailed among classical physicists 'in matters of principles: In the beginning (*if there was such a thing*) God created Newton's laws of motion together with the necessary masses and forces' (italics added). See P. A. Schilpp (ed.), *Albert Einstein: Philosopher-Scientist* (Evanston, Illinois: Library of Living Philosophers, 1949), I, p. 19.
23. 'The Evolution of the Universe', pp. 609–10.
24. P. Teilhard de Chardin, *The Phenomenon of Man*, with an Introduction by Sir Julian Huxley (New York: Harper & Brothers, 1959), p. 64.
25. E. A. Milne, *Relativity, Gravitation and World-Structure* (Oxford: Clarendon Press, 1935), p. 140.
26. E. T. Whittaker, *The Beginning and End of the World* (London: Oxford University Press, 1942), p. 4.
27. *Ibid.*, p. 63.
28. E. T. Whittaker, *Space and Spirit: Theories of the Universe and the Arguments for the Existence of God* (Hinsdale, Illinois: Henry Regnery, 1948), p. 114.
29. See the pamphlet, 'The Proofs for the Existence of God in the Light of Modern Natural Science. Address of Pope Pius XII to the Pontifical Academy of Sciences' (Washington, D.C.: National Catholic Welfare Conference, n.d.), p. 17.
30. Hoyle's disparaging utterances on theology in general and Christianity in particular are too well known to need to be documented.
31. The rate, which can also be expressed as one hydrogen atom per litre of space in every thousand billion years (see H. Bondi, *Cosmology* [2nd ed.;

Cambridge: University Press, 1961], p. 143), indicates, if it does anything, both the exceedingly low mean density of matter in the universe and the 'relative slowness' of its expansion at any given neighbourhood.

32. See his Address quoted above, p. 11.
33. Bondi, *Cosmology*, p. 144. Bondi's failure to refer to the Creator in connection with an alleged emergence out of nothing speaks for itself.
34. As exemplified by the symposium on modern theories of the structure of the universe broadcast by the BBC in late 1959 and published under the title, *Rival Theories of Cosmology* (London: Oxford University Press, 1960). H. Bondi, W. B. Bonnor, R. A. Lyttleton, and G. J. Whitrow were the panelists of the symposium.
35. See *TIME*, October 30, 1978, p. 108.
36. An example is the illustrated report in *TIME*, March 4, 1974, p. 74, which showed J. P. Ostriker, astrophysicist at Princeton University, pointing to the outer arms of a spiral galaxy in which he hoped to find that extra mass, needed, according to him, if Kepler's Third Law ruled its rotation. For further details, see my article, 'The History of Science and the Idea of an Oscillating Universe', in W. Yourgrau and A. D. Breck (eds.), *Cosmology, History and Theology* (New York: Plenum Press, 1977), pp. 233–51, and the Postscript to the new enlarged edition of my *Science and Creation: From Eternal Cycles to an Oscillating Universe* (Edinburgh: Scottish Academic Press, 1980).
37. See *The New York Times*, Nov. 2, 1975, p. 56, col. 2. Five years later and after intensive search for the missing mass, perhaps 20 per cent of its required amount can be seen as having been located, as I was told by Prof. P. J. E. Peebles of Princeton University, a leading authority on this topic, who himself is very sympathetic to the idea of an oscillating universe. His survey of the latest research concerning that mass will be published in the Lecture Notes for the 1979 Les Houches Summer School, (University of Grenoble), under the title, 'Masses of Galaxies and Clusters of Galaxies'. As one would expect, banner headlines greeted, at the end of April 1980, when this book was being typeset, C. L. Cowan's latest neutrino experiments, which may assure the existence of the missing mass, provided they stand up to independent scrutiny, a rather long process involving many theoretical and experimental difficulties.
38. S. Weinberg, *The First Three Minutes: A Modern View of the Origin of the Universe* (London: Andre Deutsch, 1977), pp. 3 and 153–5.
39. R. Jastrow, *God and the Astronomers* (New York: W. W. Norton, 1978).
40. See his letter of November 22, 1876, to the Bishop of Gloucester, in L. Campbell and W. Garnett, *The Life of James Clerk Maxwell* (London: Macmillan, 1882), pp. 382–6.

41. W. R. Inge, *God and the Astronomers* (London: Longmans, Green & Co., 1933).
42. According to Jastrow (*op. cit.*, p. 112) 'Eddington wrote in 1931: 'I have no axe to grind in this discussion, . . . the notion of a beginning is repugnant to me . . . I simply do not believe that the present order of things started off with a bang . . . the expanding Universe is preposterous . . . incredible . . . *it leaves me cold*'. Eddington's *The Nature of the Physical World* (p. 85) contains the passage: 'I simply do not believe that the present order of things started off with a bang'. He stated in his contribution to the panel discussion in 1931: 'The theory of the expanding universe is in some respects so preposterous that we naturally hesitate before committing ourselves to it. It contains elements apparently so incredible that I feel almost an indignation that anyone should believe in it – except myself' (*loc. cit.*, p. 588). Some of the remainder of the words put by Jastrow into Eddington's mouth can be found in his *New Pathways in Physical Science* (1935) as referred to in note 8 above. Jastrow's fancy seems to be the source of the words 'it leaves me cold'. Neither these words, nor the other passages can be found in articles published by Eddington in 1931, two of which touch directly on the expansion of the universe.
43. Jastrow, *God and the Astronomers*, p. 113.
44. The procedure of Morrison should seem all the more glaring because Babbage used in his book, *The Ninth Bridgewater Treatise*, his theory of calculating machines, as a pointer to the existence of God. For details, see my *Brain, Mind and Computers* (1969; new ed.; South Bend, Indiana: Gateway, 1978), pp. 43–5.
45. Jastrow, *God and the Astronomers*, p. 113. Prof. Morrison's feeling should be understandable to all those who heard him, as the second Bronowski-lecturer, argue that given enough time termites could come up with a telescope worthy of Palomar.
46. C. Sagan, *Broca's Brain: Reflections on the Romance of Science* (New York: Random House, 1979), p. 65.
47. See G. Gamow, *The Creation of the Universe* (new rev. ed; New York: The Viking Press, 1961), p. [vii].

Chapter Two

1. A. France, *La vie littéraire* (Paris: C. Levy, 1888–92), III, p. 212. The phrase is from France's review of Flammarion's *Uranie*.

2. France therefore concluded in his short story, 'Abeille',: 'Science is inhumane. It is not science but poetry that charms and consoles. This is why poetry is more necessary than science'. See A. France, *Balthasar* (Paris: C. Levy, n.d.), p. 256.
3. See Einstein's 'Autobiographical Notes', in *Albert Einstein: Philosopher-Scientist* edited by P. A. Schilpp (Evanston, Illinois: Library of Living Philosophers, 1949), I, p. 9.
4. P. A. M. Dirac, 'The Evolution of the Physicist's Picture of Nature', *Scientific American* 208 (May 1963), p. 47. This detail is not mentioned in the recent, more systematic, but in places rather esoteric treatment of the topic by S. Chandrasekhar, 'Beauty and the Quest for Beauty in Science', *Physics Today* (July 1979), pp. 25–30.
5. See ch. 8, 'The Myth of One Island', in my *The Milky Way: An Elusive Road for Science* (New York: Science History Publications, 1972).
6. The plane was foot-pedalled by Bryan Allen from Dover to Cap Gris Nez. See report, 'The Albatross Flies', in *Newsweek*, June 25, 1979, p. 97.
7. On Galileo's advocacy of 'circular inertia', see W. R. Shea, *Galileo's Intellectual Revolution: Middle Period, 1610–1632* (New York: Science History Publications, 1972), pp. 167–8.
8. The trust which Newton's authority generated in the existence of the ether was memorably put in the Preface of *A Dissertation on the Aether of Sir Isaac Newton* (London: Printed for Charles Hitch, 1747) by Bryan Robinson, who declared: 'The Evidence from the Phaenomena is much stronger in Favour of the Existence of the *Aether*, than it is in Favour of the Existence of the Air'.
9. The analogy, first proposed by G. Stokes in 1845, became, a generation later, a hallowed shibboleth through the authority of Lord Kelvin. See my *The Relevance of Physics* (Chicago: University of Chicago Press, 1966), p. 80.
10. For the great variety of experiments proposed and carried out, see E. T. Whittaker, *A History of the Theories of Aether and Electricity* (London: Nelson, 1951), I, pp. 388–92 and II, pp. 28–9. The most telling expression of Michelson's sentiments can be found in the description of the courses offered in the Physics Department, of which he was the head, of the University of Chicago. Its *Annual Register* carried from its issues of 1897–98 to 1906–07 the statement: 'While it is never safe to affirm that the future of Physical Science has no marvels in store even more astonishing than those of the past, it seems probable that most of the grand underlying principles have been firmly established and that further advances are to be sought chiefly in the rigorous application of these principles to all the phenomena which come under our notice'.
11. From a distance of some 50 years Einstein recalled his first encounter, at

the age of 16, with Maxwell's equations as a 'revelation', 'the most fascinating subject at the time I was a student'. See 'Autobiographical Notes', p. 33.

12. See, for instance, the Presidential Address, 'The Primary Concepts of Physics', of William F. Magie, Professor of Physics at Princeton University, at the meeting of the American Physical Society in Washington, D.C., on December 28, 1911; *Science* 35 (1912), pp. 281–93.
13. For an English translation of this paper, 'Cosmological Considerations on the General Theory of Relativity', see *The Principle of Relativity* by H. A. Lorentz *et al.* (1923; New York: Dover, n.d.), pp. 177–88.
14. It received wide and undisputed currency through Addison's *Spectator* (July 9, 1714) and Voltaire's *Élémens de la philosophie de Neuton.*
15. For historical and technical details, see my 'Das Gravitations-Paradoxon des unendlichen Universums', *Südhoffs Archiv* 63 (1979), pp. 105–22; on its optical counterpart, my monograph, *The Paradox of Olbers' Paradox* (New York: Herder & Herder, 1969).
16. See *The Principle of Relativity,* p. 179.
17. H. Shapley, *The View from a Distant Star: Man's Future in the Universe* (New York: Basic Books, 1963), p. 46.
18. The story, which includes Dicke, Gamow, Alpher and others, is well told in S. Weinberg, *The First Three Minutes* (London: Andre Deutsch, 1977), pp. 122–31.
19. See *TIME,* October 30, 1978, p. 108.
20. Even such baryons as the protons are not absolutely stable. Such is at least the consequence of the unification of the electromagnetic and weak forces with the force binding the nucleus of atoms together, a theory which is now being subjected to experimental test on a grandiose scale in a salt mine near Cleveland.
21. A sufficiently simplified account of that 'cooking' is given in B. Lovell's *In the Centre of Immensities* (London: Hutchinson, 1978), pp. 99–105.
22. Implicit in their technical papers, but very explicit in Hoyle's sundry utterances and in H. Bondi's *Cosmology.*
23. The claim of the steady-state theorists that their cosmological approach supports the 'perfect cosmological principle', is an unwitting echo of Aristotle's claim in his *On the Heavens,* that the universe must be eternal because it is perfect. In both cases the claim echoes unabashed paganism and involves patent contradictions. Assertion of the 'perfection' of the world implies, even in its moderate Leibnizian form of 'the best possible world', an elimination of God's freedom to create.
24. See, for instance, B. Carter, 'Large Number Coincidences and the Anthropic Principle in Cosmology', in M.S. Longair (ed.), *Confrontation*

of Cosmological Theories with Observational Data (Dordrecht: D. Reidel, 1974), pp. 291–8.

25. An infinite duration of physical processes cannot be had even in an oscillating universe, because at each bounce its starting temperature has to increase. Thus, the present expansion of the universe would not be possible if it had been preceded by an infinite number of oscillations.
26. See B. Lovell, *The Impact of Modern Astronomy on the Problems of the Origins of Life and the Cosmos* (Condon Lectures, Oregon State System of Higher Education: Eugene, Oregon., 1963), pp. 43–4.
27. Such was at least the conclusion, still undisputed, of I.D. Novikov and Ya. B. Zeldovich, 'Physical Processes near Cosmological Singularities', *Annual Review of Astronomy and Astrophysics* 11 (1973), pp. 387–410.
28. An unnamed British historian of astronomy, quoted in G. F. Becker, 'Kant as a Natural Philosopher', *American Journal of Science* 5 (1898), pp. 97.
29. Its first appearance came in the introduction of Laplace's *Théorie analytique des probabilités* (1812) and gained wide currency through its inclusion in his *Essai philosophique sur les probabilités* (1814), a non-mathematical presentation of the same topic.
30. Or of those, with however a great scientific acumen, who fail to perceive that the theoretical inability to measure atomic interactions with complete precision does not prove the absence of exact interactions.
31. For a still unsurpassed popular exposition of Eddington's 'cosmology *a priori*', see ch. xi in J. Singh, *Great Ideas and Theories of Modern Cosmology* (1961; New York: Dover, n.d.).
32. See Oppenheimer's Condon Lecture, *The Constitution of Matter* (Eugene: Oregon State System of Higher Education, 1956), p. 2.
33. Quoted in G. Holton, 'The Mainsprings of Discovery – The Great Tradition', *Encounter* (April 1974), p. 91.
34. The claim is expressed in a slightly veiled form in the printed version of his lecture, 'What Are the Building Blocks of Matter?' in D. Huff and O. Prewett (eds.), *The Nature of the Physical Universe: 1976 Nobel Conference* (New York: John Wiley, 1979), pp. 27–45; see especially p. 45.
35. See John Wilhelm's report, 'A Singular Man', *Quest* (April 1979), p. 38. Efforts aimed at an *a priori* specification of the shape and structure of the universe were also very visible at the rise of modern science. While Descartes' utterances to that effect are well remembered, different is the case with Galileo's remark in his First Letter on Sunspots (1612): 'They [epicycles etc.] are not retained by philosophical astronomers, who, going beyond the demand that they somehow save the appearances, seek to investigate the true constitution of the universe – the most important and most admirable problem that there is. For such a constitution exists; it is

unique, true, real, *and could not possibly be otherwise*; and the greatness and nobility of this problem entitle it to be placed foremost among all questions capable of theoretical solution' (Italics added). Quoted from *Discoveries and Opinions of Galileo*, translated with an introduction and notes by S. Drake (Garden City, N.Y.: Doubleday, 1957), p. 97. This is why Galileo misled himself into assuming a 'circular inertia'. See note 8 above.

36. 'Every reader of medieval Latin texts knows that few Bible verses are so often quoted and alluded to as the phrase from the Wisdom of Solomon, 11:21, "*omnia in mensura, numero et pondere disposuisti*"', was noted by E. R. Curtius in his magisterial monograph, *European Literature and the Latin Middle Ages*, translated from the German by W. R. Trask (London: Routledge and Kegan Paul, 1953), p. 504.
37. St. Augustine, *Confessions*, Bk. XI, ch. 12.
38. For details, see chapters 1 and 2 in my *The Relevance of Physics*.
39. The experiment, performed in the summer of 1979, consisted in the head-on collision of electron and positron beams at 15 billion electron volts. At such a high energy, there appears, in addition to two streams of quarks and antiquarks, respectively, also a third stream interpreted as consisting of gluons, the exchange particles which hold together the quarks, the constitutive parts of protons and neutrons.
40. See K. R. Popper, *Conjectures and Refutations: The Growth of Scientific Knowledge* (New York: Harper and Row, 1968), p. 270.
41. A glance at the name and subject index of Carnap's *The Logical Structure of the World – Pseudoproblems in Philosophy*, translated by R. A. George (Berkeley: University of California Press, 1967), which contains no such entries as Einstein, universe, cosmology, and the like, should reveal enough of solipsism, the inevitable outcome of the crusade of logical positivists against 'pseudoproblems' in philosophy.
42. E. Nagel and J. R. Newman, *Gödel's Proof* (New York: New York University Press, 1958), pp. 100–1.
43. The book in question is my *The Relevance of Physics*. See especially, pp. 127–30.
44. *American Scientist* 55 (1967), p. 352.
45. A statement made by Einstein in 1920 during a lecture given in Prague which was attended by young H. Feigl who recalled it in his 'Beyond Peaceful Coexistence', *Historical and Philosophical Perspectives of Science*, edited by R. H. Stuewer (Minneapolis: University of Minnesota Press, 1970), p. 9.
46. A. Einstein, *Lettres à Maurice Solovine* reproduits en facsimile et traduits en français (Paris: Gauthier-Villars, 1956), pp. 102–3.
47. *Ibid.*, pp. 114–15.

Chapter Three

1. About this motto, embroidered upon the Chair of State of Mary, Queen of Scots, Nicholas White wrote, following his visit with the captive Queen in the spring of 1569, to Cecil: 'In looking upon her cloth of estate, I noticed this sentence embroidered "*En ma fin est mon commencement*" which is a riddle I understand not.' Quoted in *In My End Is My Beginning*, by Maurice Baring (Preface). For the reverse of the motto, 'In my beginning is my end', see *East Coker* by T. S. Elliot.
2. 2 Mac 7:28 This and subsequent scriptural quotations are from the Jerusalem Bible.
3. The context is the recital of vicissitudes and persecutions touched off by Heliodorus' profanation of the Temple.
4. For an annotated English translation, see A. Heidel, *The Babylonian Genesis* (2nd ed.; Chicago: University of Chicago Press).
5. See especially Tablet IV, *ibid.*, pp. 36–43.
6. The somewhat dated interpretation of Yahweh as 'He who makes things happen', recently urged by W. H. Brownlee ('The Ineffable Name of God', *Bulletin of American Schools for Oriental Research* 226 [1977], pp. 39–46) is a rather transparent effort to distract from the significance of 'is' with a resort to a mere 'happening'.
7. A phrase of E. Gilson, *The Spirit of Medieval Philosophy*, tr. A. H. C. Downes (New York: Charles Scribner's Sons, 1940), p. 51.
8. A course fraught with obvious pitfalls, as amply demonstrated in the case of C. P. Snow, who after receiving on March 31, 1969, an honorary degree at the New York School of Hebrew Union College-Jewish Institute of Religion, declared that 'whatever kind of human excellence you examine', you find that 'the Jewish performance has been not only disproportionate, but almost ridiculously disproportionate', a record which is 'quite outside any sort of statistical probabilities'. While not discounting the possibility that such excellence may have been occasioned by environmental pressure, he looked for the true cause in genetics: 'Is there something in the Jewish gene-pool which produces talent on quite a different scale from, say, the Anglo-Saxon gene-pool? I am prepared to believe that this may be so.' With flagrant inconsistency, though with understandable political expediency, Sir Charles advised caution concerning A. Jensen's conclusion that intelligence is largely hereditary: 'The findings [of Jensen] should not be dramatized until there is absolute

scientific justification. I wish that Jensen had been a little more careful'. (*The New York Times*, April 1, 1969, p. 37, cols. 3–4).

9. On the 47 uses of *bārā'* in the Old Testament, see the book-length article 'Création' by H. Pinard (*Dictionnaire de Théologie Catholique*, IV, cols. 2034–2201, especially col. 2043) still unsurpassed for its richness of documentation.
10. Only by ignoring the unanimous verdict of modern biblical scholarship that the author of Genesis did not mean to present a cosmogony, can one conclude as F. Hoyle does: 'The cosmology of the ancient Hebrews is only the merest daub compared with the sweeping grandeur of the picture revealed by modern science.' *The Nature of the Universe* (1950; New York: The New American Library, 1955), p. 121.
11. It first appears in Exodus (4:22) and, with the exception of a passage in Malachi (2:10), it conveys not so much authority as compassion and forgiveness.
12. Mt 11:25.
13. Although their blame is smaller than that of those carving and worshipping idols, 'they are not to be excused: if they are capable of acquiring enough knowledge to be able to investigate the world, how have they been so slow to find its Master?' (Wisd 13:9). This passage is distinctly echoed in Paul's Epistle to the Romans (1:20), 'ever since God created the world his everlasting power and deity – however invisible – have been there for the mind to see in the things he has made', a fact which should rather discredit the slighting of the Book of Wisdom as being non-inspired on account of its alleged 'Hellenisation' of genuine 'Hebraic' revelation.
14. 'He is the image of the unseen God and the first-born of all creation, for in him were created all things in heaven and on earth: everything visible and everything invisible . . . all things were created through him and for him' (Col 1:15–17).
15. 'In the beginning was the Word . . . He was with God in the beginning. Through him all things came to be, not one thing had its being but through him' (Jn 1:1–3).
16. For the points made by Celsus as emphasized in the remainder of this paragraph, see *Origen: Contra Celsum*, translated with an introduction and notes by H. Chadwick (Cambridge: University Press, 1965), IV, 14–18 (pp. 193–5) and V, 26, 33, 41 (pp. 283, 289, 296). The many excerpts from Celsus' work, no longer extant, which are contained in Origen's refutation of it, permit a fair reconstruction of Celsus' line of argumentation. For a very readable example, see B. Pick, 'The Attack of Celsus on Christianity', *The Monist* 21 (1911), pp. 223–66.
17. See edition by H. Rabe, *Ioannes Philoponus de aeternitate mundi contra*

Proclum (Leipzig: B. G. Teubner, 1899), pp. 113–14. The length of the work is obviously a reason that it has not been translated into any of the major modern languages. The Latin translation, very rare and now more than four hundred years old, by Jean Mahot (Lyons, 1557), has not yet lost its usefulness.

18. For texts and interpretation, see my *Science and Creation: From Eternal Cycles to an Oscillating Universe* (Edinburgh: Scottish Academic Press, 1974), pp. 172–80.
19. Compilers of *The Oxford English Dictionary* (1933) must have noted with a touch of despair that 'genesis' can mean *any* mode of formation. See vol. IV F–G, p. 109.
20. The contrast in *Timaeus* (37b) between the world of generations (coming-into-being) and the immutable (heavenly) realm weakens the portent of Plato's surprising view that time was generated, a view which Aristotle found incompatible with the eternity or ungenerated character of the universe.
21. What makes this first appearance of the expression '*ex nihilo*' all the more remarkable is Tertullian's insistance that 'it is a norm of faith (*regula fidei*) . . . that there is only one God and there is no other God beside the creator of the world who produced everything out of nothing (*de nihilo*) through his Word'. *De praescriptione haereticorum*, c. xiii (PL II, 26). For this and other texts mentioned in this paragraph, see Pinard's article quoted above.
22. *De animae procreatione*, 1014b.
23. *Metaphysics*, 1075b.
24. *De coelo*, 298b.
25. *De Melisso, Xenophane et Gorgia disputationes*, 975a.
26. A. Ehrhardt, 'Creatio ex Nihilo', *Studia theologica* (Lund) 4 (1951), p. 24. Ehrhardt's reasoning displays its true nature from the very first paragraph of his article, as he declares that the biblical phrase 'in the beginning' and the expression '*creatio ex nihilo*' are largely contradictory, and counsels a 'very necessary reserve towards the more scientific language which asserts that God created the world out of nothing' (p. 13). His characterization of '*ex nihilo*' as a scientific and not a philosophical idea is, on the one hand, confusing (science never deals with 'nothing') and, on the other, reveals something of Ehrhardt's resolve to present Bible and Christian creed as rigorously a-philosophical.
27. A. Ehrhardt, *The Beginning: A Study in the Greek Philosophical Approach to the Concept of Creation from Anaximander to St. John*, with a memoir by J. Heywood Thomas (Manchester: University Press, 1968), p. 167. For the passage from Atticus, see p. 166.
28. Some of these arguments are listed by St. Augustine in his *On the City of God*, Bk. 11, ch. 4–6 and Bk. 12, ch. 15–16.

29. From Julian's 'Hymn to King Helios', in *The Works of the Emperor Julian*, with an English translation by W. C. Wright (London: W. Heinemann, 1913–23), I, p. 399.
30. The more favourable light, in which Philo's ideas on creation are put by H. A Wolfson in his *Philo: Foundations of Religious Philosophy in Judaism, Christianity, and Islam* (Cambridge, Mass.: Harvard University Press, 1947), should be balanced with the evaluation in E. Bréhier, *Les idées philosophiques et religieuses de Philon d'Alexandrie* (Paris: A. Picard, 1907).
31. According to the article 'Creation' in *The Jewish Encyclopedia* (New York: Funk & Wagnalls, 1903, IV, pp. 336–40), early Rabbinic literature conveys the impression that decision on the notions of creation *ex nihilo* and in time 'is of no consequence to the practical religiosity which Judaism means to foster', that from the earliest times 'there were many among the Jews who spoke out on this subject with perfect candor and freedom' and that the eternity of matter 'had many adherents among medieval Jews'. As to modern Judaism, the article 'Creation' in *Encyclopedia Judaica* (New York: Macmillan, 1971, V, pp. 1059–71) comes to a close with an endorsement of the philosophy of 'emergence' savouring of pantheism: 'The moral implication of the traditional teaching that God created the world is that creativity, or the continuous emergence of aspects of life not prepared for or determined by the past, constitutes the most divine phase of reality'. In view of these rather representative accounts of the Jewish idea of creation, the allegation of H. Jonas, that the Christian dogmas of Incarnation and Trinity divested it of its pristine genuineness, will appear rather contrived. See his 'Jewish and Christian Elements in the Western Philosophical Tradition', in D. O'Connor and F. Oakley (eds.), *Creation: The Impact of an Idea* (New York: Charles Scribner's Sons, 1969), pp. 241–58, see especially, pp. 250–1.
32. The most complete collection and interpretation of those remarks is still R. T. Herford's *Christianity in Talmud and Midrash* (London: Williams and Norgate, 1903), a work whose scholarly merit was not disputed by J. Klausner, author of *Jesus of Nazareth: His Life, Times and Teaching* (translated from the original Hebrew by H. Danby; New York: Macmillan, 1926), which has remained a classic among books on Jesus by modern Jewish authors. Klausner reported (p. 48) that although many modern printings of the Talmud do not contain those anti-Jesus passages, these have nonetheless lived on as an oral tradition to be recalled especially around Christmas. It must, however, be noted that many among the Reformed Jews have developed a high esteem towards Jesus in recent times.
33. According to that conciliar definition, aimed mainly at the Albigenses, 'We firmly believe and plainly confess that the true God is One . . . who

through His omnipotent power created out of nothing at the beginning of time simultaneously both creatures, spiritual and corporeal, that is, angelic and of this world'.

34. A. Ehrhardt, *art. cit.*, p. 41.
35. On the fate of Greek science among the Arabs, see my *Science and Creation*, ch. 9.
36. The remark in *The Jewish Encyclopedia* (IV, p. 339) that 'Maimonides is most timid in his defense of creation', tells more than extensive documentations could convey, especially when taken in conjunction with the words in *Encyclopedia Judaica* (V, p. 1068) that 'with respect to the question whether creation occurred *ex nihilo* Maimonides claims that this issue is not crucial for religious faith'.
37. See E. Gilson, *The Spirit of Medieval Philosophy*, pp. 433–4.
38. For the provenance of this statement of Ockham and for related utterances of his, see my *The Road of Science and the Ways to God* (Chicago: University of Chicago Press, 1978), pp. 41–3.
39. F. Oakley, 'Christian Theology and the Newtonian Science. The Rise of the Concept of the Laws of Nature', in D. O'Connor and F. Oakley (eds.), *Creation: The Impact of an Idea*, pp. 54–83; see especially, pp. 64–5. Particularly difficult, within this Ockhamist-Protestant perspective, is the explanation of the role which Copernicanism played in the rise of modern science not only through the mediation of Galileo's Christian Platonism, but especially through the crucial role of Kepler's Three Laws as transmitted by Horrocks to Newton.
40. Epoch-making is indeed the phrase in Buridan's commentary on Aristotle's *De caelo: 'Posset enim dici quod quando deus creavit sphaeras coelestes, ipse incipit movere unamquamque earum sicut voluit; et tunc ab impetu quem dedit eis, moventur adhuc, quia ille impetus non corrumpitur, cum non habeant resistentiam*'. See *Johannis Buridani Quaestiones super libris quattuor de caelo et mundo,* edited by E. A. Moody (Cambridge, Mass.: The Medieval Academy of America, 1942), pp. 180–1. This phrase, repeated almost word for word by Oresme, Buridan's foremost disciple, and also professor at the University of Paris, became a standard feature of late medieval commentaries on Aristotelian cosmology, and an indisputable source for the subsequent formulation of such fundamental notions of Newtonian physics as momentum and inertia.
41. The next four paragraphs are a summary of arguments developed and documented in chapters 3–8 of my *The Road of Science and the Ways to God.*
42. Unfortunately, this quotation is one of the very few, whose provenance is not given by A. O. Lovejoy in his *The Great Chain of Being: A Study of the History of an Idea* (1936; New York: Harper and Row, 1960), p. 150. According to D. Saurat, Professor of French literature at the University of

London, Hugo's espousal of pantheism was a result of his close connection between 1836 and 1852 with Alexandre Weill, a well-known Jewish expert in cabbalistic lore (*La religion de Victor Hugo* [Paris: Hachette, 1929], p. 19).

43. Since the emotional approach, when skilfully presented, has a powerful appeal, it is never popular to take immediately a critical view of it. An ever modern pattern is indeed illustrated in the fact that it took years before due credit was given to the perspicacity of that Oxford don, who, on hearing H. L. Mansel buttress, in his Bampton Lectures of 1859, religion by setting it aside from critical objective reasoning, was prompted to remark: 'I had not expected to live to hear atheism preached from the pulpit of the University'. See V. F. Storr, *The Development of English Theology in the Nineteenth Century 1800–1860* (London: Longmans, Green and Co., 1913), p. 422.
44. Quoted after Dom Cuthbert Butler, *The Vatican Council: The Story from Inside in Bishop Ullathorne's Letters* (London: Longmans, Green and Co., 1930), II, p. 253. The official annotations to this definition make it all too clear that it was through the phrase 'from the very first beginning of time' that pantheism was to be struck at its very root. Consequently the phrase must be seen as an integral specification of the sense in which the dogma should be understood.

Chapter Four

1. *Dialogue concerning the Two Chief World Systems – Ptolemaic and Copernican*, translated by Stillman Drake (Berkeley: University of California Press, 1962), p. 53. See also p. 230, where Galileo claims the absence of rigorous (geometrical) proofs 'in the whole ordinary philosophy'.
2. Popper's theory of three worlds was unwittingly discredited by Popper himself through his programmatic warning that there was no room for the knowing subject in his epistemology. Most readers of Popper's *Objective Knowledge* (Oxford: Clarendon Press, 1974), where this was stated (pp. 106–7), obviously failed to ponder how the proponent of the theory of three worlds could himself know it unless he was a 'knowing subject' and how others could objectively know this.
3. *Tractatus Logico-Philosophicus*, 1.1. See German text with an English

translation, with an introduction by B. Russell (London: Kegan Paul, Trench, Trübner, 1922), p. 31.

4. The opening phrase of Russell's introduction to the *Tractatus*, 'whether or not it prove to give the ultimate truth on the matters with which it deals' (p. 7), should be seen in the light of Russell's fear that Wittgenstein may have demonstrated conclusively that all mathematics was tautology. See note 28 below. Wittgenstein felt wholly confident: 'I believe, I've solved our problems finally. This may sound arrogant but I can't help believing it', he wrote to Russell on March 10, 1919. See *Letters to Russell, Keynes, and Moore*, edited by G. H. von Wright (London: B. Blackwell, 1974), p. 67.
5. M. Ward, *Return to Chesterton* (New York: Sheed and Ward, 1952), p. 153.
6. N. Malcolm, *Ludwig Wittgenstein: A Memoir*, with a biographical sketch by G. H. von Wright (London: Oxford University Press, 1958), p. 70.
7. *Ibid.*, p. 69.
8. *Tractatus Logico-Philosophicus*, 1, p. 31.
9. B. Russell, *A Critical Exposition of the Philosophy of Leibniz* (Cambridge: University Press, 1900), see especially ch. 15 on the proofs of the existence of God.
10. See *Leibniz Selections*, edited by P. P. Wiener (New York: Charles Scribner's Sons, 1951), pp. 345–6.
11. *Ethica*, Part I, prop. xvii, see *The Chief Works of Benedict de Spinoza*, translated with an introduction by R. H. M. Elves (1883; New York: Dover, 1955), II, p. 386.
12. In a letter to Tschirnhausen, *ibid.*, II, p. 409.
13. Although this phrase, quoted in E. Gilson's *Le réalisme méthodique* (Paris: P. Téqui, n.d. [1937], p. 41), is not indicated in the two massive volumes of *Lexicon Spinozanum* compiled by E. G. Boscherini (The Hague: M. Nijhoff, 1970) Gilson's customary carefulness with historical material seems to be a sufficient guarantee for its authenticity.
14. Indeed, it put Locke on the road to solipsism, an outcome of which something was unwittingly indicated by Locke himself. Witness his *Essay concerning Human Understanding* in which he states that the mind can know only ideas and that all of man's ideas 'are within his own breast, invisible and hidden from others' (Bk III, ch 2, §1, see also Bk IV, ch. 1, §1).
15. A revealing sign of the futility of Hume's reasoning aimed at abolishing objective causality, is that its impact on the world of philosophy can only be stated by implying such causality. A case in point is the encomium heaped by V. Mehta on Wittgenstein and Hume: 'Wittgenstein's earthquake hit the philosophers of the twentieth century as hard as David Hume's cyclone – which swept away cause and effect from the human

experience – had hit their eighteenth-century predecessors'. *Fly and the Fly-Bottle* (Harmondsworth: Penguin, 1965), p. 83.

16. This hopeless effort, which at least on Fichte's part went contrary to Kant's express disavowal, deserves a comment only because its patently disastrous consequences did not dissuade J. Maréchal from trying it again. It tells something of the inexorable though rather slow (in most cases) working of logic that only a few saw a petard being hoisted on the gate of Christian philosophy when Maréchal declared the philosophies of Fichte, Schelling, and Hegel to be not only cases of Transcendental Idealism but also of genuine metaphysics (*Le point de départ de la métaphysique. Cahier V, Le thomisme devant la philosophie critique* [Paris: Felix Alcan, 1926], p. 436). It was in such terms that he hoped to restore the pristine purity of Kantianism (Kant notwithstanding) and the appeal of Thomism (self-styled Thomists applauding).
17. *History and Root of the Principle of the Conservation of Energy*, translated and annotated by P. E. B. Jourdain (Chicago: Open Court Publishing Company, 1911), p. 54.
18. For a survey of that theory, see M. Jammer, *The Philosophy of Quantum Mechanics: The Interpretations of Quantum Mechanics in Historical Perspective* (New York: John Wiley and Sons, 1974), pp. 507–21. For a criticism of Jammer's survey, see my *The Road of Science and the Ways to God* (Chicago: University of Chicago Press, 1978), pp. 411–12.
19. In a letter of December 22, 1950, to Schrödinger. See *Letters on Wave Mechanics: Schrödinger, Planck, Einstein, Lorentz*, edited by K. Przibram, translated with an introduction by M. J. Klein (New York: Philosophical Library, 1967), p. 36.
20. See E. Gilson, *Réalisme thomiste et critique de la connaissance* (Paris: J. Vrin, 1939), pp. 9–40.
21. See 'Account of the Life and Writings of Thomas Reid' (1803) in *Thomas Reid: Philosophical Works*, with notes and supplementary dissertation by Sir William Hamilton (reprint of the 8th edition, Edinburgh, 1895, with an introduction by H. M. Bracken; Hildesheim: Georg Olms, 1967), I. p. 28.
22. In a letter quoted by F. Mentré, 'Pierre Duhem, le théoricien 1861-1916', *Revue de philosophie* 29 (1922), pp. 457–8.
23. Texts quoted in Gilson, *Réalisme thomiste et critique de la connaissance*, pp. 20–1.
24. *Summa theol.* I, 87, 3 *resp.*
25. G. K. Chesterton, *Saint Thomas Aquinas* (Garden City, N.Y.: Doubleday, 1956), p. 185.
26. Such is the gist of the grand, concluding question, 'Does it really help to imagine that there is some one full, objective, true account of nature and

that the proper measure of scientific achievement is the extent to which it brings us closer to that ultimate goal?' in *The Structure of Scientific Revolutions* (Chicago: University of Chicago Press, 1962, p. 170), by T. S. Kuhn, who also boasts of the fact that the word 'truth' has not appeared in his book until its very last pages. But if the existence and success of science can be accounted for, as Kuhn wants it, merely 'in terms of evolution from the community's state of knowledge at any given time', then Kuhn's book too is subject to the consequences of the same severance of knowledge from objectivity and should be considered to be at best an expression of Kuhn's state of mind fully conditioned by others' thinking about science, and in no way an objective critique of their science of an objective reality.

27. As memorably argued by E. Meyerson in his *Identity and Reality* (translated by K. Loewenberg; 1930; New York: Dover, 1962).

28. Russell himself provides the information (*Introduction to Mathematical Philosophy* [London: George Allen and Unwin, 1919, p. 205] that he began to suspect the tautological status of mathematical and logical propositions under the influence of Wittgenstein. The publication of Wittgenstein's *Tractatus* turned suspicions into an unpalatable truth, as Russell himself recalled from a distance of almost half a century in his letter of December 20, 1968, to Prof. C. W. K. Mundle: 'I did not appreciate that his work implied a linguistic philosophy. When I did we parted company . . . I felt a violent repulsion to the suggestion that "all mathematics is tautology". I came to believe this but I did not like it. I thought that mathematics was a splendid edifice, but this shows that it was built on sand'. Quoted in R. W. Clark, *The Life of Bertrand Russell* (London: Jonathan Cape and Weidenfeld & Nicolson, 1975), p. 370. 'All mathematical proof', Russell wrote around 1951, 'consists merely in saying in other words part or the whole of what is said in the premises'. *Essays in Analysis*, edited by D. Lackey (New York: George Braziller, 1973), pp. 304–5.

29. For a documentation of this pathetic phase of scientific thought, see my *The Milky Way: An Elusive Road for Science* (New York: Science History Publications, 1972), especially ch. 1, 4 and 5.

30. A. Noyes, *Watchers of the Sky* (New York: Frederick A. Stokes, 1922), pp. 226–27. These marvellous lines sound rather ironical as put into the mouth of Newton, a champion of proofs of God's existence based on the gaps of scientific knowledge.

31. Words of Blainville, emphatically endorsed by Comte in his *Cours de philosophie positive,* III (Paris: Bachelier, 1838), p. 625.

32. Jean-Paul Sartre, *Nausea,* translated by Lloyd Alexander, with an

introduction by H. Carruth (New York: New Directions Publishing Company, 1959), p. 177.

33. *The Scientific Outlook* (1931; New York: W. W. Norton, 1962), p. 95.
34. He did so in his Herbert Spencer Lecture, 'On Scientific Method in Philosophy', in *Mysticism and Logic* (Garden City, N.Y.: Doubleday, 1957), p. 94.
35. *Nausea*, p. 176.
36. S. Hampshire, 'Metaphysical Systems', in D. F. Pears (ed.), *The Nature of Metaphysics* (London: Macmillan, 1957), p. 23.
37. Curiously, those circles are invariably sympathetic to the writings of such rationalists as J. J. C. Smart whose admission of the inconclusiveness of Kant's criticism of the cosmological argument received wide publicity in *The Cosmological Argument: A Spectrum of Opinion*, edited by D. A. Burrill (Garden City, N.Y.: Doubleday, 1967, p. 267).
38. J. J. C. Smart, 'The Existence of God', (1955), *ibid.*, p. 278.
39. *Philosophy* (New York: W. W. Norton, 1927), p. 14.
40. Letter to Hugo Bell (1674) in *The Chief Works of Benedict de Spinoza*, II, p. 386.
41. The expression is from *Le témoignage du sens intime opposé à la foi profane et ridicule des fatalistes modernes* (Auxerre: F. Fournier, 1760), II, p. 259, by J.-A. Le Large de Lignac (1710–62), best remembered for his penetrating criticism of Lockean empiricism in his *Éléments de métaphysique tirés de l'expérience* (1753), which rivals some of the best works of Scottish common-sense philosophers.
42. The cosmologist was A. V. Ambartsumian, of the Soviet Academy of Sciences; the conference the 16th World Congress of Philosophy, Düsseldorf; his statement was made at the plenary session for cosmology, in the morning of August 28, 1978.
43. See Clarke's Boyle Lectures (1704), *A Demonstration of the Being and Attributes of God* (7th ed.; London: James and John Kapton, 1728), p. 49. Further details are given in my *The Road of Science and the Ways to God*, pp. 91–92.
44. Most of those 36 proofs in Mersenne's *Quaestiones celeberrimae in Genesim* (1624) are based on geometrical patterns and their apparent purposefulness. See my Gifford Lectures, *The Road of Science and the Ways to God*, pp. 62–4. As I have noted there, the debilitating weakness of Mersenne's natural theology, a result of his empiricism, evidenced itself, twenty years later, in a complete despair about the possibility of proving God's existence. It was the same empiricism, which prompted him to argue, around 1635, that it was impossible for man to discover any true, that is, exact, law of nature, because God could have created things in any other way (see his 'Nouvelles observations physiques et mathématiques', in his

Harmonie universelle [1636]; reprinted, Paris: Centre National de la Recherche Scientifique, 1963, III, p. 8). Tellingly, Mersenne had in mind Galileo's claim about the times-squared law of free fall which turned out to be an exact law. The attitudes of Mersenne and Galileo illustrate two extremes, both of which are refuted by science. Despair about the abilities of the human mind (and a concomitant exaggeration of cosmic contingency) is as unscientific as is an overestimate of those abilities (accompanied with a slighting of cosmic contingency), a posture exemplified by Galileo (see ch. 2, note 35).

45. For a reproduction and discussion, see my *Planets and Planetarians: A History of Theories of the Origin of Plenatary Systems* (Edinburgh: Scottish Academic Press, 1978), p. 87.
46. Y. Gaillard, *Buffon. Biographie imaginaire et réelle. Suivie de Voyage à Montbard par Hérault de Séchelles.* Preface d'Edgar Faure (Paris: Hermann, 1977), pp. 158–9.
47. G. K. Chesterton, *Orthodoxy* (1909; Garden City, N.Y.: Doubleday, 1959), p. 149.

Chapter Five

1. C. W. Misner, K. S. Thorne, J. A. Wheeler, *Gravitation* (San Francisco: W. H. Freeman, 1973), p. [v].
2. In a conversation with M. Clynes, during the winter 1952–53. See P. Michelmore, *Einstein: Profile of the Man* (New York: Dodd, 1962), p. 251.
3. March 15, 1971, p. 6, cols. 3–6.
4. E. Haeckel, *Freedom in Science and Teaching*, translated from the German, with a Prefatory Note by T. H. Huxley (New York: D. Appleton, 1879), pp. 92–93.
5. See Marx's letters to Engels, December 19, 1860 and January 16, 1861, in their *Selected Correspondence 1846–1895*, with explanatory notes, translated by D. Torr (New York: International Publishers, 1942), pp. 125–6.
6. A first-hand report on this is given by V. L. Kellogg, a biologist serving with the American relief mission at the German Great Headquarters in occupied France in 1915, in his *Human Life as a Biologist Sees It* (New York: H. Holt, 1922), p. 51.
7. D. Gasman, *The Scientific Origins of National Socialism: Social Darwinism in Ernst Haeckel and the German Monist League* (London: Macdonald, 1971).

8. London: Constable, 1919, p. xiii.
9. This incompleteness of understanding is even greater concerning the rise of genera, and grows exponentially as the emergence of even more encompassing units – classes, orders, and phyla – is considered.
10. A. Huxley: *Ends and Means: An Inquiry into the Nature of Ideals and into the Methods Employed for Their Realization* (New York: Harper and Brothers, 1937), p. 316.
11. Not only the supernatural but also a purely natural Providence may seem to be pre-empted by the same train of thought. Witness a famous letter of Rousseau, generally considered to be the most concise formulation of his philosophical creed, which he wrote on August 18, 1756, to Voltaire. While he could consider himself to be of more worth, in the eyes of God, than an entire planet, he could not see why he 'should weigh more, in His eyes, than all the inhabitants of Saturn', whose existence he held to be most probable. (*The Complete Works of Voltaire*, vol. 101 [Banbury: The Voltaire Foundation, 1971], pp. 286–7). Beneath Rousseau's reasoning, and despair, there lay his reliance on the principle of plenitude which assured him of the spontaneous emergence of intelligent beings. Clearly, instead of stressing the (purely hypothetical) possibility of Incarnation on other planets, Christian theologians should have centred their reply on the irreducibility of mind to matter, a theme which can hardly be handled by theologians distrustful (and ignorant) of a realist metaphysics.
12. Thus, for instance, W. Penfield: 'For myself, after a professional lifetime spent in trying to discover how the brain accounts for the mind, it comes as a surprise now to discover, during this final examination of the evidence, that the dualist hypothesis seems the more reasonable of the two possible explanations'. *The Mystery of the Mind* (Princeton, N.J.: Princeton University Press, 1975), p. 85.
13. Sherrington's 'dualistic philosophy' is mentioned almost with an apology by Judith P. Swazey in her article 'Sherrington' in *Dictionary of Scientific Biography* vol. xv (New York: Charles Scribner's Sons, 1975), p. 401. Swazey's article may have been completed before the publication in 1970 of *Facing Reality* (New York: Springer Verlag, 1970), by J. C. Eccles, who quotes (p. 174) from his conversation with Sherrington a week before the latter's death on March 4, 1952. The conversation contains the following statement of Sherrington: 'For me now the only reality is the human soul'. See also J. C. Eccles and W. C. Gibson, *Sherrington: His Life and Thought* (New York: Springer International, 1979), pp. 182–3.
14. C. Sagan, *Broca's Brain: Reflections on the Romance of Science* (New York: Random House, 1979), pp. 3-12. The sight of Paul Broca's brain in the Musée de l'Homme prompted Sagan to reflections which are both crudely materialistic and unscientific: 'Is Paul Broca still there in his formalin-

filled bottle? Perhaps the memory traces have decayed, although there is good evidence from modern brain investigations that a given memory is redundantly stored in many different places in the brain. Might it be possible at some future time, when neurophysiology has advanced substantially, to reconstruct the memories or insights of someone long dead?' (p. 9).

15. In an unsigned review, 'Geological Climates and the Origin of Species', of two books of Sir Charles Lyell, in *The Quarterly Review* 126 (1869), pp. 359–94; for quotation see p. 392. Typically, Sagan, who recalls Wallace as a co-proponent of the theory of evolution by natural selection (*op. cit.*, pp. 129–30), fails to recall Wallace's insistence on the inability of that theory to cope with the reality of consciousness.
16. The remark of T. Edinger, of Harvard, made at the International Colloquium on Paleontology, held in Paris, April 1955, still retains the validity of its gripping concreteness: 'If man passed through a phase pithecanthropus-sinanthropus, the evolution of his brain was unique not only in its result, but also in its tempo. An increase of 50 to 100 percent of the cerebral hemisphere is a phenomenon that occurred also in the Equides, between the stages representated by Merychippus, of medium size, and Equus, of large size. In the Equides this transformation took twenty-five million years. With the Hominides, the same increase seems to have taken place, geologically speaking, in an instant without having been accompanied by a major increase of size.' Translated from his address in French, 'Objets et résultats de la paléoneurologie', *Annales de paléontologie* 42 (1956), p. 99.
17. *The Origin of Species* (6th ed.; London: John Murray, 1876), p. 265.
18. He did so in his Presidential Address to the Geological Society in 1870; see *Lay Sermons, Addresses and Reviews* (New York: D. Appleton, 1871), p. 244.
19. *The Evolution of Man: A Popular Exposition of the Principal Points of Human Ontogeny and Philogeny,* translated from the German (New York: D. Appleton, 1892), I. p. xxiii.
20. The investigator in question is N. J. Berrill, formerly Professor of zoology at McGill University. Quotations are from his *The Origin of Vertebrates* (Oxford: Clarendon Press, 1955), pp. 10 and 248.
21. In his letter of March 29, 1863, to J. D. Hooker; see F. Darwin, *The Life and Letters of Charles Darwin* (London: John Murray, 1888), III, p. 18.
22. In that formula, as given in C. Sagan (ed.), *Communication with Extraterrestrial Intelligence* (Cambridge, Massachussetts: MIT Press, 1973), $R_\star$ stands for the rate of star formation; f_p for the fraction of stars with planetary systems; n_e for the mean number of planets within such systems that can be abodes of life; f_l for the fraction of planets on which life

actually occurred; f_i for the fraction of planets with intelligent life; f_c for the fraction of planets with advanced technological civilizations; and L for the mean lifetime of such civilizations.

23. W. H. McCrea, 'Astronomer's Luck', *Quarterly Journal of the Royal Astronomical Society* 13 (1972), pp. 506–19; see especially pp. 517–18.
24. For details and documentation, see my *Planets and Planetarians: A History of Theories of the Origin of Planetary Systems* (Edinburgh: Scottish Academic Press, 1978).
25. Or to recall the caustic remark of W. W. Howells, a past president of the American Anthropological Association: 'I will lay a small bet that the first men from Outer Space will be neither bipeds nor quadrupeds but bimanous quadrupedal hexapods. I have just invented the last word, in the hope that it means six limbs'. *Mankind in the Making* (Garden City, N.Y.: Doubleday, 1959), p. 345.
26. A central theme of the 'Symposium on the Implications of Our Failure to Observe Extraterrestrials' held at the University of Maryland, November 2–3, 1979.
27. C. N. Yang, M. Ryle, for instance. Advances in manned space flight during the last two decades have left intact the appropriateness of the conclusion of another Nobel laureate, E. Purcell, according to whom plans to travel to another planetary system belong to where they are mostly advertised, the sides of cereal boxes. 'Radioastronomy and Communication through Space', Brookhaven Lecture Series, Number 1, November 16, 1960, p. 11.
28. *Biology and the Exploration of Mars*, edited by C. S. Pittendrigh and others (Washington, D.C.: National Academy of Sciences Research Council, 1966), p. 8.
29. September 16, 1969, p. 46. 'No Life on Mars'.
30. Quoted in *TIME*, September 20, 1976, p. 87.
31. Quoted in *TIME*, September 3, 1979, p. 63.
32. He did so on more than one occasion. See *The Scientific Papers of Sir William Herschel*, edited by J. L. E. Dreyer (London: The Royal Society, 1912), I, p. 479 and II, p. 147.
33. See Lambert's *Cosmological Letters on the Arrangement of the World-Edifice*, translated with an introduction and notes by S. L. Jaki (New York: Science History Publications, 1976), p. 83.
34. See S. L. Jaki, 'An English Translation of the Third Part of Kant's *Universal Natural History and Theory of the Heavens*', in W. Yourgrau and A. D. Breck (eds.), *Cosmology, History and Theology* (New York: Plenum Press, 1977), pp. 387–403, and my translation of the entire work with introduction and notes (Edinburgh: Scottish Academic Press).
35. It was a supreme irony that Bruno's *La cena de le ceneri* (see my translation,

The Ash Wednesday Supper, with introduction and notes [The Hague: Mouton, 1975]) happened to be the first book on Copernicus. For as Frances A. Yates, whose *Giordano Bruno and the Hermetic Tradition* (Chicago: University of Chicago Press, 1964) revolutionized studies on Bruno, remarked: 'Copernicus might well have bought up and destroyed all copies of the *Cena* had he been alive' (p. 297).

36. A. O. Lovejoy, *The Great Chain of Being: A Study of the History of an Idea* (1936; New York: Harper and Row, 1960), p. 332.
37. See J. M. Pasachoff, *Contemporary Astronomy* (Philadelphia: W. B. Saunders, 1977), pp. 424–5.
38. See S. S. Kumar, 'Planetary Systems', in W. C. Saslaw and K. C. Jacobs (eds.), *The Emerging Universe* (Charlottesville: University Press of Virginia, 1972), pp. 25–34.
39. L. Eiseley, *The Immense Journey* (New York: Vintage Books, 1959), p. 162.
40. The next three paragraphs owe much to McCrea's article quoted in note 23 above.
41. Discussed in some detail in my Fremantle Lectures, *The Origin of Science and the Science of Its Origin* (Edinburgh: Scottish Academic Press, 1978), pp. 15–6.
42. For further details, see *ibid.*, pp. 31–7.
43. The first five volumes were published in 1913–17 (Paris: Hermann). Duhem's sudden death in 1916 prevented the immediate publication of the last five volumes which saw print only in the 1950s.
44. He did so at the C. H. Boehringer Sohn symposium (Kronberg, Taunus, 16–17 May 1974) on *The Creative Process in Science and Medicine*, edited by H. A. Krebs and J. S. Shelley (Amsterdam: Excerpta Medica, 1975), p. 126.
45. This is a paraphrase of a remark of Chesterton in his preface to *Cosmology* by D. C. O'Grady (Ottawa: Graphic Publishers, 1932), p. [16].
46. As reported by J. Monod at the Boehringer Sohn symposium; see *The Creative Process in Science and Medicine*, p. 123.
47. *Science and Civilisation in China* (Cambridge: University Press, 1954–), II, p. 581.
48. *Maitri Upanishad*, First Prapathaka, in *Thirteen Principal Upanishads*, translated by R. H. Hume (2nd rev. ed; London: Oxford University Press, 1934), p. 414. For further details see my *Science and Creation: From Eternal Cycles to an Oscillating Universe* (Edinburgh: Scottish Academic Press, 1974), pp. 7–9.
49. Such is the grand conclusion, hardly an improvement on Rousseau, in J. Goodfield, *Playing God: Genetic Engineering and the Manipulation of Life* (London: Sphere Books, 1978), p. 240.

INDEX OF NAMES